When seen hunting over a meadow, barn owls have an ethereal grace and beauty that can be matched by no other bird. The barn owl has an almost global distribution and has lived in close proximity to humans since settlement and farming created places in which to nest and the forest clearings needed for hunting. However, in many countries, barn owl numbers are falling rapidly. This book explores the relationship between barn owls and their prey worldwide and demonstrates how an understanding of such relationships can help in the conservation of the species. In this comprehensive account, Iain Taylor describes the biology and ecology of this species, including the factors affecting breeding success and causes of mortality that influence the final recruitment of new birds into the population. He concludes by suggesting ways in which we can manage and conserve this beautiful bird for the future.

This book will be of interest to researchers, professionals and interested amateurs in the fields of ornithology, ecology, natural history and conservation.

Barn owls

Predator–prey relationships and conservation

BARN OWLS

Predator – prey relationships and conservation

IAIN TAYLOR
University of Edinburgh

CAMBRIDGE
UNIVERSITY PRESS

Published by the Press Syndicate of the University of Cambridge
The Pitt Building, Trumpington Street, Cambridge CB2 1RP
40 West 20th Street, New York, NY 10011-4211, USA
10 Stamford Road, Oakleigh, Melbourne 3166, Australia

First published 1994

Printed in Great Britain at the University Press, Cambridge

A catalogue record for this book is available from the British Library

Library of Congress cataloging in publication data

Taylor, Iain.
    Barn Owls : predator–prey relationships and conservation / Iain
Taylor.
        p.   cm.
    Includes bibliographical references and index.
    ISBN 0-521-39290-X
    1. Barn owl.   2. Predation (Biology).   3. Birds, Protection of.
I. Title
QL696.S85T38   1994
598.9'7 – dc20   93-26509 CIP

ISBN 0 521 39290 X hardback

RO

To the memory of Tom Irving, of Langholm

Contents

Preface

Predatory animals hold a special fascination for many people. Compared with herbivores, they must spend a high proportion of their time searching for their food and, once discovered, prey have to be captured and subdued before being eaten. This has often led to dramatic anatomical and behavioural adaptations which from a human viewpoint can be spectacular. Few could fail to be impressed by the graceful running of the cheetah in pursuit of antelopes, or the awesome speed of a peregrine falcon swooping towards a pigeon. As symbols of power and majesty, predators have long held a special place in human society and folklore.

For natural historians and ecologists, predators, and their relationships with their prey, have been a source of great inspiration. As far back as 1927, Charles Elton in his classic book *Animal Ecology*, which was one of the first truly scientific attempts to explore nature, discussed the relationships between arctic lemmings and arctic foxes and between the snowshoe hare and the lynx. He was able to call upon long runs of quantitative information which had been collected by fur trappers and these revealed the existence of distinct population cycles in both predator and prey: a phenomenon which despite years of subsequent research is still not fully understood. Another of the early ecologists Paul Errington explored the relationships between many predators and their prey but especially between mink and muskrats in the marshlands of Iowa, a study which emphasised the need to consider the animals' social behaviour (Errington 1943). Authors such as these laid the foundations of the scientific study of predators; a study which has blossomed into all aspects of behaviour, physiology, energetics, genetics and population and community ecology.

In more recent times, predators have attracted attention for other reasons. Because of their position at the top of food chains and their need for large areas of land in which to live, they have suffered more than most other animals from man's exploitation and pollution of the natural environment. It was through the devastating effects on populations of predators that we were first alerted to the menace of organochlorine pesticides. Predators are our natural early warning systems and, even if for no reason other than selfishness, we should conserve them. Their loss would be our loss, sooner or later.

In this book I have tried to pursue these two themes: the relationship between barn owls and their prey, and the conservation of barn owls. The two are entirely complementary; without a scientific understanding of the owl's ecology and behaviour we cannot hope to guarantee their conservation. Barn owls may lack the power and speed of the more dramatic predators but for what they lack in these departments they more than make up for in grace and beauty. When seen hunting over a meadow in the misty sunrise of a summer morning they have an almost ethereal quality that can be matched by no other bird. There are few dissenters in the argument to conserve these wonderful creatures.

For the ecologist they offer a veritable feast of fascination. My own interest in them began when I was living in Africa where they were common birds, but my studies in Scotland started in earnest in 1978 and stretched for 15 years up until 1992 and these studies form the basis of much of the book. I have, however, also attempted to place these observations within the context of other studies to try to achieve a more global account of the species. I have therefore drawn freely on the published, and in some cases unpublished, work of others and to all I owe an enormous debt of gratitude. Many, I have had the pleasure to meet and to be shown around their study areas. I hope I have done justice to their work.

My own researches were done in the south of Scotland and in some respects they were the result of team work. Tom Irving, to whose remembrance this book is dedicated, lived in the town of Langholm in my main study area and was the local link with most of the farmers, all of whom knew and respected Tom. He accompanied me on fieldwork whenever his work permitted, which he ensured was often. Tom cared deeply for all the local fauna but barn owls were special for him. We had many happy days together in the field and Tom's untimely death was a devastating loss to us all.

Many others helped for short or long periods, but in particular it is a pleasure to thank Ian Langford who participated in the project for several years. Murdo Macdonald, Aleem Chaudhary, Peter Bell, Mairi Osborne, Fiona Slack, Donna Kreft, Murray Grant, Dominic McCafferty, Laurel Hanna and many undergraduate students helped in numerous ways. Mick Marquiss introduced me to the art of radio-tracking and Nigel Charles taught me his method for estimating small mammal abundance. I am grateful to the many farmers who could not have been more helpful or more enthusiastic, but in particular to

Thompson and Janet McKnight of Glencartholm Farm and to Ella, Tom Irving's wife, for their endless hospitality. Buccleuch Estates and other landowners generously granted permission to work on their ground.

The list of barn owl researchers whose studies feature in the book is long indeed. Without exception they are wonderful and open people. To those who showed me their work in the field, I am particularly grateful. In France, Hugues Baudvin introduced me to the delights of church bell towers, favoured nesting places for barn owls, and gave me valuable insights into the ecology of French barn owls. Michel Juillard introduced me to the Swiss versions. I learned an immense amount from both of these totally dedicated people. David Wells and John Duckett showed me their studies in Malaysia and David generously allowed me to use his unpublished measurements of Malaysian barn owls. In New England, Bruce Colvin was unstinting in his hospitality; his studies of north American barn owls have almost exactly paralleled my own and it was fascinating to see the similarities and differences. John Cayford showed me his study area in south-east England.

I have gained enormously from discussions with others whose study areas I have been unable to see Johan de Jong in Friesland, Yves Müller in France, Hubertus Illner in Germany, Carl Marti from Utah and David Hollands from Australia. David's own recently published book on Australian owls is totally absorbing. Walter Boles of the Australian National Museum, Sydney, kindly compiled data on measurements of Australian barn owls which he has allowed me to use here. A special thanks also to the staff of the Scottish Ornithologists Club, Edinburgh whose excellent library was invaluable in researching some aspects of the book.

Over the years I have benefited from discussions with other owl and raptor researchers and these have shaped much of my thinking and inevitably the nature of this book. In particular, Ian Newton has never failed to provide inspiration. Likewise, Pertti Saurola, Erkki Korpimaki, Hannu Pietiainen, Mick Marquiss and Steve Petty have all unwittingly added to the contents of these pages. Ian Newton, Hannu Pietiainin and John Love read various parts of the book in draft and made many useful comments. Dave Haswell prepared some of the diagrams and printed my photographs and Baxter Cooper drew the diagram of barn owl foraging habitats to replace a somewhat embarrassing effort of my own from 1981. To both, I am extremely grateful. Margaret Jackson typed the manuscript. Robert Smith, Don Smith and Ken Taylor provided many of the beautiful photographs of barn owls and Johan de Jong supplied some from Holland. Keith Brockie's wonderful illustrations speak for themselves. My wife, Sarah, helped in the field on many occasions and was a constant support during the writing of the book.

Iain Taylor
December 1992
Edinburgh

1 Introduction

This book explores the relationships between barn owls and their prey and demonstrates how an understanding of such relationships can be applied to the conservation of the species. Of course, not all aspects of the owls' ecology and behaviour can be interpreted solely with reference to their food supply; there are many other fascinating facets to explore. The main themes therefore have numerous side branches and points of confluence that I hope, together, provide a reasonably comprehensive account of these wonderful birds.

Predator–prey relationships are two-way processes, but here I have concentrated on the effects of the prey on the predator. These effects ramify throughout almost all parts of the owls' lives: their habitat requirements, their breeding and moult, their movements and survival, and much else. To understand all of these, it is necessary to understand what kind of predator the barn owl is: what species it hunts, whether or not it has preferred prey that are more profitable to catch, how its diet changes through time and space with changing environmental conditions, how it finds and catches its prey and what habitats it hunts over. The birds' energetics have to be explored so that their needs at different times can be appreciated, and also the significance of factors such as weather conditions must be understood as they could affect the behaviour of the prey, the owls' ability to hunt and also their energy requirements.

Familiarity may sometimes breed contempt but understanding generates respect and this, in our battle to conserve nature, is our most powerful weapon. Hunches and guesses sometimes have their place but there are no crystal balls: we have no sure way of knowing what will come next. It is a dangerous person who tries to divorce conservation from scientific research for by doing so it may deny nature the best possible chance. Our wildlife deserves the very best we can give it. Scientific knowledge that enables us to predict how nature will respond to future changes and a caring and sympathetic public who are prepared to implement whatever is needed must be the ways of achieving our ultimate goals. I hope that this book may take us at least a few steps in that direction.

1.1 The barn owl: briefly

Most people in Europe and North America will be familiar with the barn owl as a star of calendars and Christmas cards but only those who live and work in the countryside will have much knowledge of the real bird. In life, it is quiet and unobtrusive and, for the most part, nocturnal in its habits. It is seldom seen except perhaps for a fleeting glimpse in car headlights. Mostly, in the north, it is a bird of farmlands and coastal marshes but for its origins we have to look elsewhere.

Barn owls have an almost global distribution matched by few, if any, other species and, although the fossil record is poor, it seems likely that they originated in warm savannahs or lightly wooded environments and that they expanded northwards following the spread of human settlement and farming. Even within historic times this advancement can be traced in North America with the movement of European settlers. They nest in cavities and must originally have depended upon tree holes and fissures in cliffs as well as the abandoned nests of some other species. Artificial structures, especially farm buildings, provided excellent alternatives at the same time as forest clearance created the open habitat they needed for hunting (Fig. 1.1).

Whether in their original environments or in man-created habitats, they mostly hunt small terrestrial mammals, mainly rodents, which they locate by using their sense of hearing. Their long wings and legs and

Fig. 1.1. A female barn owl delivers prey to her young in the roof space of an abandoned cottage. (Photograph: Donald Smith.)

other aspects of their anatomy demonstrate that they are adapted to hunt over open areas: short to medium length grasslands and patchy, scrub-like habitats. Their natural specialisations must have suited them well to life in farmland and until recently they were highly successful. Now, their numbers are declining and their range contracting over much, if not most, of Europe and North America.

1.2 Study areas in Scotland

I began my studies of barn owls in the summer of 1978 by searching for them in the farmlands around the city of Edinburgh. To the east of the city the fertile flat landscape had been cultivated intensively and few owls remained but to the south, in areas of poorer soil where the habitat was more varied, I found many breeding pairs (Area 4, Fig. 1.2). However, most were nesting in elm trees and this presented problems. Often there was only a small entrance into large complex cavities so that it

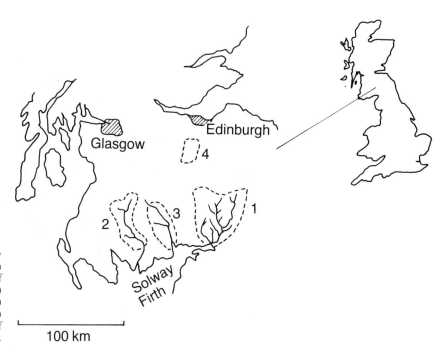

Fig. 1.2. Location of study areas in Scotland: (1) Main study area, the catchment of the Esk and Liddle rivers. (2) The Dee river catchment. (3) The Nith river catchment. (4) Part of Peeblesshire, south of Edinburgh.

was impossible to see well enough to be sure of the nest contents. At that time Dutch Elm disease was beginning to take a heavy toll of the trees and as I intended that the study should be long-term this did not bode well. I therefore moved the study to the south-west of Scotland where there were known to be a good number of owls nesting in farm buildings. On the advice of Ian Newton, I sited my main study area around the valley of the River Esk.

I was particularly anxious to include within the area all of the habitat types and altitudes that the owls were likely to use, so I decided that it should encompass the entire catchment of the River Esk, which included the River Liddle and many smaller streams. I also decided to establish a number of replicate study areas to act as a check on the wider applicability of results from the main area. I settled on the catchment of the River Dee, 75 km to the west of the Esk, and on the River Nith around Dumfries, 45 km west (Fig. 1.2).

The Esk area covered 1600 km² and rose from fertile coastal plains to uplands reaching a height of about 500 m. Below about 150 m the land was devoted predominantly to pastoral farmland for dairy, beef and sheep production. Most fields were short rotation pastures, silage and hay, with only a small amount of arable land. Throughout, there were numerous small woodlands covering about 20% of the area

Fig. 1.3. View in the Esk study area, looking north. The enclosed pastoral farmland with small woods and hedges grades quickly into conifer plantations (behind the deciduous wood) and sheepwalk (in the background). (Photograph: Iain R. Taylor.)

(Fig. 1.3). Mostly these were coniferous but there were also some deciduous and mixed woods. Rough grassland areas that could support populations of small mammals such as voles and shrews were restricted to damp hollows in some fields and along edge habitats, especially woodland edges and along farm tracks. Because of the livestock grazing, few hedgerows supported rank grassland 'headlands' at their bases; most of those that did were along roads and tracks or where cereals were grown. There were few drainage ditches but there were numerous small streams, almost always supporting narrow strips of woodland. The majority of the farm houses and 'steadings' (barns and animal housing) dated from the 18th and 19th centuries and as a result of changes in the economy of farming in the area, many were no longer in use. Likewise, many former farm workers' cottages had been abandoned. These buildings were the main nesting places for the owls.

Above about 150 m the land rapidly graded into uplands devoted to sheep production and plantation forestry. The main human activity was centred along the numerous small valleys most of which supported narrow strips of enclosed fields of pasture and hay. Either side, the land was largely unenclosed wet rough grassland, grazed at low density by sheep. Along most valleys there were strips or patches of coniferous woodland often with wide edges of ungrazed long grassland. Farm buildings and large hardwood trees in which the owls could nest were largely restricted to the valleys. For simplicity this system of open rangeland and valleys is referred to subsequently as 'sheepwalk', a traditional Scottish term, but agriculturally this area is more accurately described as 'hill margin' land.

All of these upland areas were originally covered in mixed species forests but, through the activities of charcoal burners and sheep grazing, these have long since gone. Since the 1950s, but especially in the 1970s, large tracts of this land, usually the areas of poorer quality, have been drained and planted with even-edged woodlands of exotic conifers such as the Sitka spruce (*Picea sitchensis*). Depending upon soil type these new woodlands occur at altitudes above 120 m, with most between 200 m and 250 m. They were fenced to exclude livestock and fertilised with nitrates and phosphates to encourage growth in the early stages. Before canopy closure, which was usually around 6–8 years after planting, the trees grew up among a dense rank sward of grassland which provided excellent and extensive habitat for voles and shrews (Fig. 1.4). After canopy closure, long grassland was still found around the forest edge and along fire breaks, and eventually, after about 40 years' growth, on patches that had been clearfelled (Fig. 1.5). Nesting places for the owls in this habitat were scarce, as few large hardwood trees remained. However, during the afforestation programmes, many shepherds' cottages were left abandoned and these provided good, although short-term nest sites. As these buildings are no longer maintained, they eventually lose their roofs and become unsuitable.

Climatic information for the area was obtained from two meteorological stations: Eskdalemuir, sited at the head of the Esk catchment at 240 m, and Carlisle, just south of the lowest point at 26 m. The mid altitude between these two was also the modal altitude of barn owl nest

Fig. 1.4. Before canopy closure, young plantations have a dense grass sward giving good small mammal habitat. (Photograph: Iain R. Taylor.)

Fig. 1.5. View across the Liddle valley showing extensive areas of recently clearfelled conifer plantations. (Photograph: Iain R. Taylor.)

sites during the study. The stations therefore gave the climatic extremes of the study area in relation to altitude and mean values of the two gave a good measure of weather variables applicable to the owl population. Generally, at the lowest altitudes summers were cool and winters mild with little snow or prolonged low temperatures. With increasing altitude conditions deteriorated rapidly: temperatures were consistently lower, windspeeds higher and rainfall, snow and sleet more frequent with prolonged periods of deep snow cover in most years. Throughout most of the study area night-time temperatures, when the owls did most of their hunting, were low for much of the year. At Eskdalemuir, for example, between November and April, temperatures at grass level fell below freezing on 61% of nights (Fig. 1.6).

The altitude range within which most of the barn owl nesting attempts occurred, 75–175 m, experienced average winter temperatures and days of snow cover characteristic of most of lowland Britain. Rainfall was typical of most of western Scotland, Wales and western England, but above that of central and eastern England.

The River Dee study area (Area 2, Fig. 1.2) encompassed all except the lower 10 km of the river catchment, an area of 200 km². Compared with the Esk catchment, this was a narrower valley but it contained similar habitat types with coniferous plantations at higher altitude and sheepwalk and enclosed pastoral farmland at the lowest altitudes. The climate was similar to the middle altitudes of the Esk area. The River Nith study area (Area 3, Fig. 1.2) did not enclose the entire catchment of the river but was a transect along its length from the uplands to the sea. Once again, this had similar habitats to the other areas.

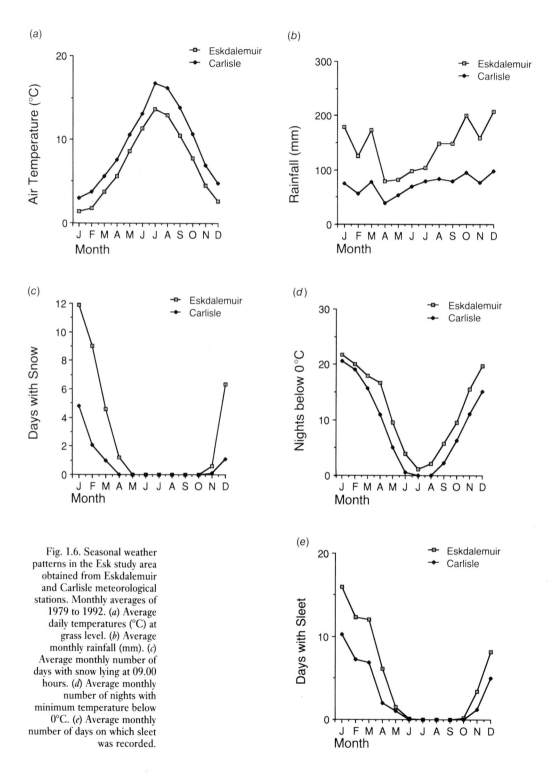

Fig. 1.6. Seasonal weather patterns in the Esk study area obtained from Eskdalemuir and Carlisle meteorological stations. Monthly averages of 1979 to 1992. (*a*) Average daily temperatures (°C) at grass level. (*b*) Average monthly rainfall (mm). (*c*) Average monthly number of days with snow lying at 09.00 hours. (*d*) Average monthly number of nights with minimum temperature below 0°C. (*e*) Average monthly number of days on which sleet was recorded.

1.3 Research methods

Ecological studies of birds of prey are nearly always a compromise between breadth and depth. Progress in our understanding can only be achieved by very detailed investigations in which precise information is obtained from well-known individuals in known circumstances. Experimental studies involving, for example, manipulation of the food supply can also be used to provide more conclusive evidence of relationships. This level of detail must be accompanied by adequate sample sizes that are representative of the population under study so that the general applicability of the results can be ensured. Replicate study areas also greatly strengthen the validity of conclusions. With owls and most diurnal raptors, obtaining good, unbiased samples is usually difficult as the birds occur at low densities and are normally hard to find. Pairs nesting in more conspicuous locations are likely to be over-represented. Seemingly impressive sample sizes, if not representative, are of less value than small but adequate and representative samples.

In my studies, I tried to obtain representative samples by using river catchments as study areas, so that all habitats and altitudes were included, and by locating all breeding pairs in these areas each year. No easy method for finding nesting pairs was discovered so intensive searches of all possible breeding places had to be carried out each year. And as the potential breeding season is long, these searches had to be done at least twice, all of which was extremely time consuming. Initially it took 140 days of fieldwork just to locate the possible nesting places such as tree holes, farm buildings and abandoned cottages, in the three study areas.

The most detailed studies were made in the main area, the Esk study area, over 14 years and here I attempted each year to catch as many of the breeding and non-breeding adults as possible as well as to record laying dates, clutch sizes, hatching success, chick growth and survival, and fledging success. As many as possible of the young were ringed each year to obtain information on dispersal, although it was never possible to ring all of them. Many of the derelict buildings in which they nested were highly unstable and, though I could see and count the chicks, any attempt to reach them could have been catastrophic for all parties. For most pairs each year I obtained data on diet from pellet analyses and from food stores at their nests. The abundance of field voles and common shrews available to the owls in the field were monitored by trapping each spring. Other field procedures such as radio-tracking, automatic recording of nest visits, habitat mapping and trapping of small mammals in different habitats were carried out over shorter time periods.

Such intensive and seemingly intrusive investigations would have been worthless if the birds had reacted adversely to them, so I took particular care in the first year of the project to ensure that they did not. I made my first visits to nests when they had large chicks and gradually advanced the visits, checking all the time on the responses of the birds. Eventually I discovered that I could catch birds at the nest even before laying with no adverse effect. The attachment of radio transmitters to their tail feathers also had no detrimental effect (Taylor 1991a). Most professional ecologists are driven as much by a love and admiration for their study animals as by scientific pursuit and it is normal to expect that the well-being of their study animals would take the highest priority. The farmers were amazed by my caution as many said they had been in the habit of visiting nests regularly to check on progress and had never noted desertions. Indeed many of the birds gradually habituated to visits and became almost tame. By contrast, I have found captive birds to be much more aggressive and nervous. Hugues Baudvin, who has studied barn owls in France for many years, has also not experienced problems with nest visits. However, in the USA, Bruce Colvin has advised strongly against visiting nests before the young are well advanced. The North American subspecies is a much more aggressive bird than the European one and this, along with the definite tendency to desert nests if visited early, may be a result of living alongside serious potential nest predators such as racoons.

Studies in the two replicate study areas continued over 7 years from 1980 to 1986 and involved fewer visits, mainly to determine the number of breeding pairs, clutch size and fledging success. Vole abundance in these areas was not quantified by trapping but a record was kept of peak and decline years from the abundance of signs such as fresh droppings and runways. The results in all aspects of the owls' ecology were similar for all areas but could not be combined as vole abundance was quantified only in the Esk area. Throughout the following chapters therefore I have concentrated on the results from the Esk area and referred to the other areas only when by doing so a better insight into particular problems is provided.

2 Distribution and variation

P resent day owls are subdivided into two families, the Tytonidae which includes the barn owls (subfamily Tytoninae) and the bay owls (subfamily Phodilinae), and the Strigidae which includes all other owls. The soft bones of birds are preserved only under special circumstances and specimens of great age are rarely discovered. As a result, the evolutionary origins of barn owls are unknown and the fossil record is too scanty to trace their development in any meaningful way. The earliest known barn owl remains, mostly fragments from the ends of leg bones, have been found in deposits from the Miocene period (25–12 million years ago) in what are now France and North America. The evidence from other fossils suggests that the climate in France at that time was considerably warmer than at present and the plant and animal communities may have been similar to that of the present day savannahs

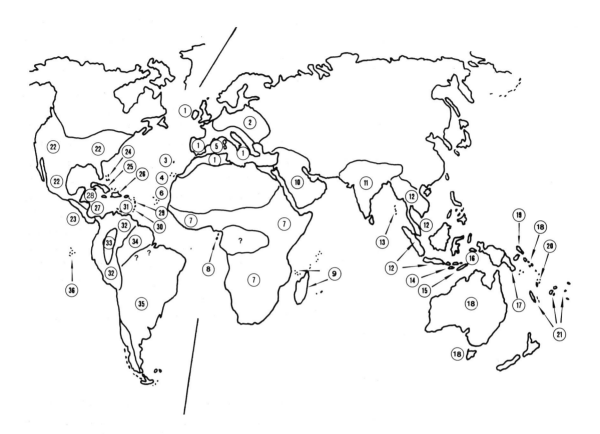

Fig. 2.1. Distribution of barn owl subspecies. 36 subspecies have been described, although some confusion exists over the status of some. Precise details of distribution are uncertain for many and the map is intended as a general guide. **1.** *T. a. alba* (Scopoli): UK, Ireland, Channel Is., Spain, Portugal, west and south France, Italy, Yugoslavia, Greece, N. Africa. **2.** *T. a. guttata* (Brehm): Denmark, Netherlands, Belgium, Germany, eastern Europe. Hybrid zone with *alba* in eastern France/western Germany. **3.** *T. a. schmitzi* (Hartert): Madeira. **4.** *T. a. gracilirostris* (Hartert): Canary Is. **5.** *T. a cinesti* (Kleinschmidt): Corsica, Sardinia. **6.** *T. a. detorta* (Hartert): Cape Verde Is. **7.** *T. a. affinis* (Blyth): Africa, south of Sahara. **8.** *T. a. thomensis* (Hartlaub): Sao Thome'. **9.** *T. a. hypermetra* (Grote): Comoros Is., Malagasi. **10.** *T. a. erlangeri* (Sclater): Saudi Arabia, Oman, Gulf states north to Lebanon, Syria, Iraq, Iran. **11.** *T. a. stertens* (Hartert): India, Pakistan, Bangladesh, Sri Lanka, Assam, Sikkim, Nepal, Bhutan, Burma. **12.** *T. a. javanica* (Gmelin): Thailand, Burma, Indo-China, Malaysia, Indonesia, Java, Flores, Timor. **13.** *T. a. deseopstorfii* (Hume): Andaman Is. **14.** *T. a. sumbaensis* (Hartert): Sumbe Is. **15.** *T. a. everetti* (Hartert): Savu Is. **16.** *T. a. kuehni* (Hartert): Lesser Sunda Is., Flores to Timor; confusion possible with distribution of *javanica* and *everetti*. **17.** *T. a. meeki* (Rothschild and Hartert): south east New Guinea, Vulcan and Dampier Is. **18.** *T. a. delicatula* (Gould): Australia, Solomon Is. **19.** *T. a. crassirostris* (Mayr): Boang Is., Tanga Group, Bismark Archipelago. **20.** *T. a. interposita* (Mayr): Santa Cruz Is., Banks Is., northern New Hebrides. **21.** *T. a. lulu* (Peale): New Caledonia, south New Hebrides, Fiji, Loyalty, Tonga, Samoa, Society Is. **22.** *T. a. pratincola* (Bonaparte): North and Central America. **23.** *T. a. guatemalae* (Ridgeway): Panama to Guatemala. **24.** *T. a. lucayana* (Riley): Bahama Is. **25.** *T. a. furcata* (Temminck): Cuba. **26.** *T. a. niveicauda* (Parkes and Phillips): Is. of Pines, Cuba. **27.** *T. a. bondi* (Parkes and Phillips): Bay Is. (off Honduras). **28.** *T. a. glaucops* (Kaup): Tortuga and Hispaniola, West Indies. **29.** *T. a. nigrescens* (Lawrence): Dominica, West Indies. **30.** *T. a. insularis* (Pelzeln): Lesser Antilles. **31.** *T. a. bargeri* (Hartert): Curacao Is. (off Venezuela). **32.** *T. a. contempta* (Hartert): Columbia, Ecuador, Peru, Venezuela. **33.** *T. a. subandea* (Kelso): parts of Columbia, Ecuador. **34.** *T. a. hellmayri* (Griscom and Greenway): Guianas to Amazon. **35.** *T. a. tuidara* (Gray): Brazil (south of Amazon), Chile, Argentina. **36.** *T. a. punctatissima* (Gray): Galapagos Is.

of Africa. At least six forms or species of barn owl have been recovered from sites of Pleistocene age in the Caribbean, Europe and the Mauritius Islands (Brodkorb 1972, Mourer-Chauvire *et al.* 1980). It seems likely that by a million years or so ago, and perhaps even much earlier, the barn owl group had achieved a wide distribution. Remains clearly belonging to *Tyto alba* have been discovered in many prehistoric sites throughout Europe and the new world and, today, the species has one of the widest distributions of any bird. It is found almost everywhere except the most inhospitable desert regions, closed primary rainforests, northerly and mountainous areas with extreme winters and some remote islands. Some 35 or 36 subspecies have been identified, many of them island forms with restricted distributions (Fig. 2.1). These vary greatly in body size and plumage coloration and in aspects of their behaviour and ecology. Some, such as the Galapagos barn owl *T. alba punctatissima* are so distinct and isolated that an argument could easily be made for full species status. However it is not the purpose of this chapter to delve into taxonomic details. Rather, it is an attempt to examine the variation in the barn owl group, both within and amongst subspecies, from an ecological perspective.

2.1 Variation within subspecies

Barn owls are generally medium to small-sized owls and most subspecies have long wings and legs that enable them to hunt by slowly quartering suitable open habitats and to dive into long vegetation to catch their prey. Some are adapted to more wooded environments and have shorter wings. Their heart-shaped facial disc is very well developed and is a characteristic of the species. The precise significance of these features and the way they are used by the owls are discussed later (Chapter 4).

The barn owl's soft plumage and detailed patterns are not easily captured by the written word. Their general coloration varies according to subspecies from deep grey and rich buff to soft golden washes and immaculate white; but always, on these backgrounds, there are the most intricate and delicate markings in whites, greys and black (Fig. 2.2). In most, and perhaps all, subspecies the females are much darker than the males with deep buff and grey on the upperparts and often with buff on the underparts as well. They tend to have much darker and larger spots on wing and tail feathers which gives a distinctly barred effect in flight. Also, on their underparts, they usually have larger and more numerous black flecks (see Fig. 4.6, p. 55).

In the *alba* subspecies in Scotland, most males of breeding-age were perfectly white over all of their underparts, the white colour extending

Fig. 2.2. The long wings and intricately marked upperparts of a female *alba*. Note the pronounced wing bars. (Photograph: Iain R. Taylor.)

up the sides of their neck and facial disc (Fig. 2.3). Most females had at least a wash of light ochre and many were quite richly coloured below. At the extreme end of the variation, a few males had almost no buff on their upperparts, the grey and black markings contrasting sharply with the white background, but most were suffused with a golden rather than buff coloration. Females were mostly buff or a rich golden ochre above and strongly marked with grey and black. Wing and tail bars were more pronounced in females and some males lacked these entirely (see Fig. 8.7, p. 118) so that in flight their primaries appeared totally white.

There was a subtle change in plumage with age: at fledging 44% of the males had a faint buff wash over their chest and the females were much darker underneath than older birds. By breeding age, the following year, most males had lost all trace of this coloration with only 12% retaining slight buff smudges on the sides of the chest, but whether this occurs by moulting or fading is unknown. Females continued to become paler throughout their lives so that some very old individuals of about seven years or more were as pale as some males.

Within any pair of owls it was usually easy to separate males and females on the basis of coloration even when seen at some distance (Fig. 2.4), but the most reliable character for separating captured birds in the hand at all times of the year was the amount of black flecking on the underparts. Almost all females (98%, $n = 182$) had at least some flecking on the underbody contour feathers and underwing coverts. The

Fig. 2.3. An immaculately white, unflecked male *alba* brings prey into his nest in an old barn. (Photograph: Robert T. Smith.)

Fig. 2.4. Male and female barn owl (*alba* subspecies) at the nest. The much paler plumage of the male is seen clearly. (Photograph: Kenneth Taylor.)

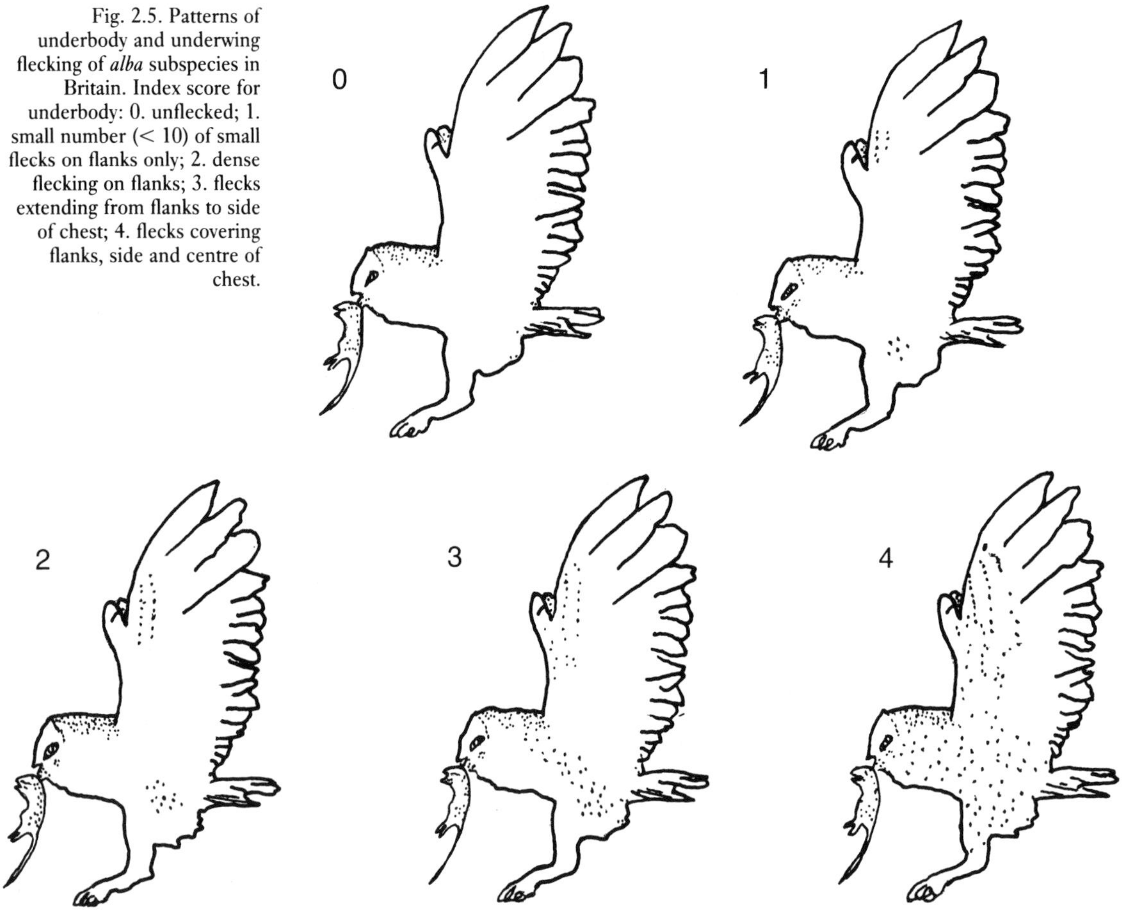

Fig. 2.5. Patterns of underbody and underwing flecking of *alba* subspecies in Britain. Index score for underbody: 0. unflecked; 1. small number (< 10) of small flecks on flanks only; 2. dense flecking on flanks; 3. flecks extending from flanks to side of chest; 4. flecks covering flanks, side and centre of chest.

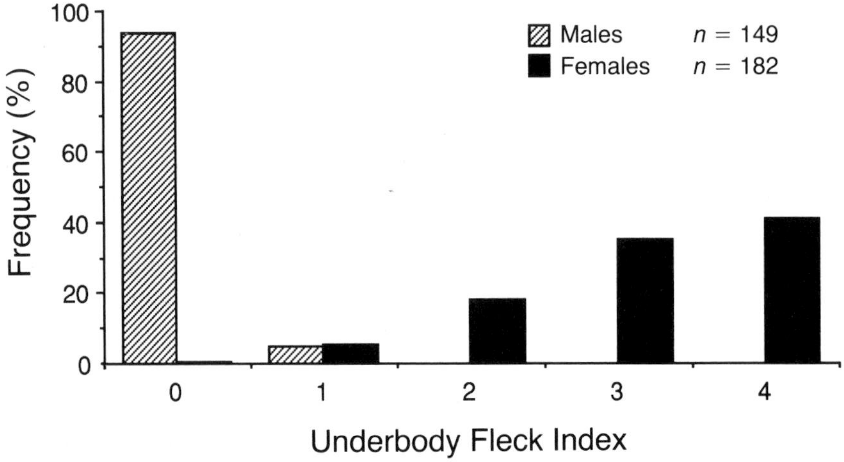

Fig. 2.6. Underbody fleck scores of breeding-age males and females, south Scotland. Most males are unmarked with scores of 0; most females have at least some flecks and the majority are strongly flecked with scores of 3 and 4.

least heavily marked had only a few small flecks on their flanks and, at the other extreme, many had their entire underparts densely covered in large diamond-shaped flecks (Figs. 2.5, 2.6). By contrast 95% ($n =$ 149) of males had no markings on their underbody and 98% had no underwing flecks. Those few that were marked had only a small number of thin, often pencil-line flecks (Figs. 2.5, 2.6). These differences between adult males and females might not be directly applicable to newly fledged birds. In these, there was a much higher incidence of more heavily flecked individuals than in the adult breeding population suggesting either that there was a skewed sex ratio in favour of females or that some males had plumages more similar to adult females at fledging. Work is still in progress to resolve this. However, all newly fledged birds that lacked flecks proved to be males when subsequently recaptured as breeding adults and all those with fleck scores of three or four proved to be females, so most of the young could be sexed accurately.

Few other subspecies exhibit such clear plumage differences between the sexes. Males of the North American subspecies are lighter coloured than their females but have much more numerous and larger underbody flecks than males of nominate subspecies. Nevertheless, both the density and diameter of these flecks are significantly less in the males than in the females and a high proportion of birds can be separated on plumage characters (Colvin 1984, Looman 1985). These general trends seem also to apply to many other subspecies including the dark-breasted European *guttata* and those in Africa and Australia.

Linear measurements of the nominate subspecies in Scotland, including wing, tail and tarsus lengths did not differ significantly between sexes (Appendix 1). A small number of females attained slightly greater wing lengths than the males (Fig. 2.7). However, most of these measurements were taken during the breeding season when females were less active than males (Chapter 11) and when some had moulted in new flight feathers earlier than males (Chapter 8), so the differences could have been explained entirely by females having less worn and newer feathers. The body weights of females varied seasonally, reaching a peak at laying. Throughout the whole breeding season, breeding females were much heavier than males (Fig. 2.8). Body weight alone was usually sufficient to distinguish the sexes of known pairs at this time, but as the females always had at least some trace of a brood patch and males do not develop this feature (Howell 1964) assessment by weighing was usually unnecessary. Non-breeding females had weights only slightly above and overlapping that of males in summer and the same was true of females in general in winter, so that weights, without a knowledge of

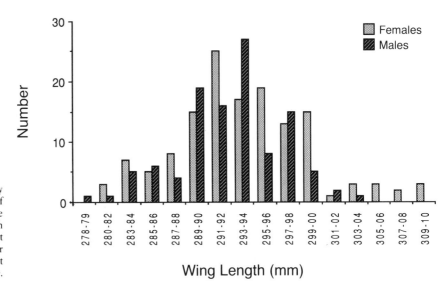

Fig. 2.7. Frequency distribution of wing lengths of breeding male and female barn owls (*alba*), south Scotland. The sexes do not differ significantly in this or any other linear measurement of body size.

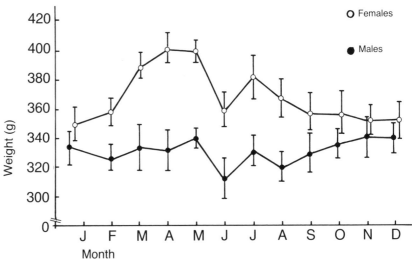

Fig. 2.8. Seasonal weight changes of breeding, adult male and female barn owls, south Scotland (mean monthly values ± standard deviations). Male weights change little throughout the year whereas female weights increase before laying and decrease steadily during incubation and brooding. The weights of both sexes tend to fall slightly when feeding the young. These data are for the high vole years of 1980 and 1981: in years of low vole abundance, the spring increase in female weights occurs up to 6 or 7 weeks later.

the status of the bird, could often be an unreliable indicator of sex. The significance of these body weight changes and differences are discussed later (Chapters 9 and 11). Similar seasonal body weight trends and differences between sexes have been described for the North American subspecies (Colvin 1984, Marti 1990) and females also tended to be a few millimetres longer in their wing, tail and tarsus measurements than males, but the differences were either not statistically significant or only weakly so (Marti 1990).

For individual males and females in the Scottish population, I could find no correlation among longest toe length, tarsus length, wing and tail lengths and body weight during the breeding season. Those with the longest wings or tarsi did not tend to be heavier and those with the longest tarsi did not have the longest toes or wings. Individual birds simply varied in the relative proportions of different parts of their body and this has implications when attempts are made to assess the 'size' of a bird. Often wing length is taken as an indicator of size but these results suggest that in barn owls it is a somewhat arbitrary index.

The most reliable method I have found for determining the age of birds after they have fledged from the nest is based upon the sequence of moult of the primary feathers (see Chapter 8). In their first moult, which occurs during and after their first breeding season (in their second calendar year of life) most birds replace only a single primary P6. Up until that time, all of the primaries are of the same age and when their undersides are examined under good light they are of the same degree of whiteness. After the replacement of P6 and until the next moult, all of the primaries look distinctly off-white other than P6 which shows up clearly. Thus breeding birds caught at the nest can readily be identified as first years and second years on the basis of this difference. At their second moult they normally replace between two and four primaries either side of P6 and at the next moult the remainder are renewed, so that birds that are third-year breeders or older have several new feathers showing whiter than the others. I tested this method using ringed birds of known age and wrongly assessed the age of only two out of a sample of 157 individuals. The same method has been developed completely independently for North American birds (Bruce Colvin, personal communication).

2.2 Variation among subspecies

I have been able to track down measurements for 22 of the subspecies of the barn owl, although for many the data are incomplete (Appendix 1). Considerable variation in body weight and in the linear measurement of various anatomical features are apparent. Body weights are available for only six subspecies and comparisons are best made among males as they show little seasonal variation, so that any seasonal biases in the collection of the original data that were not specified would have little effect. The heaviest by far are those of the south-east Asian subspecies *javanica* in which males averaged 555 g, followed by the North American subspecies at 474 g (Fig. 2.9). The two European subspecies, and the African and Australian forms, are rather similar and much lighter at around 300–330 g.

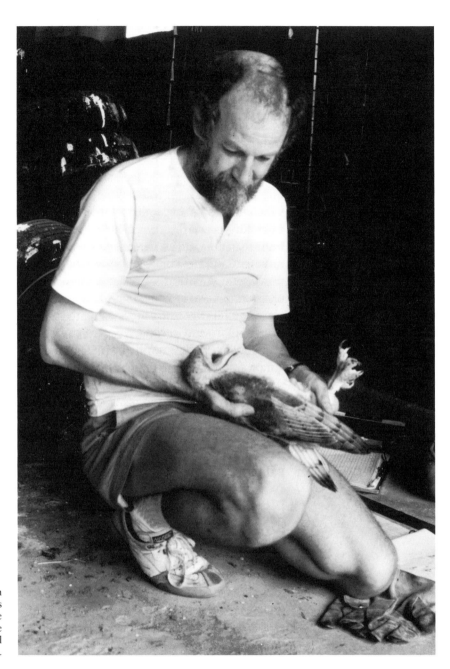

Fig. 2.9. The North American subspecies *pratincola* is considerably larger than the European subspecies. Note the robust legs and powerful talons.

Among these six subspecies, the south-east Asian form has distinctly shorter wings in relation to its body weight. This subspecies is well-known in Malaysia in oil palm plantations, which it has recently colonised, but little is known of its natural habitat. However, it probably evolved in a more forested environment than most other subspecies and hunts mainly by scanning from perches under the canopy (Chapter 4) so shorter wings would be a distinct advantage. Some of the South American subspecies also seem to have relatively short wings so the same argument may also apply to them.

Compared with the two European forms, the African, Australian, Indian and South American *tuidara* subspecies seem to have longer tarsi in relation to other aspects of their anatomy. The Australian form, for example, has a tarsus length some 23% greater than that of the nominate race even though its wing length and body weight are somewhat less. Longer legs would be an advantage when catching prey in longer vegetation so this may be an adaptation to life in tropical and sub-tropical grasslands which tend to be taller that those of temperate areas. The truly specialist grassland owl species of these habitats such as the grass owl, *Tyto longimembris*, have extremely long legs, hence the scientific name. I was unable to find much published information on toe or talon lengths but from photographs it seems that these characters also vary among subspecies. A detailed analysis would probably be rewarding.

The most marked general difference in size among subspecies is between those inhabiting small islands and those of continents and large islands (Fig. 2.10, Appendix 1). Island forms have significantly shorter

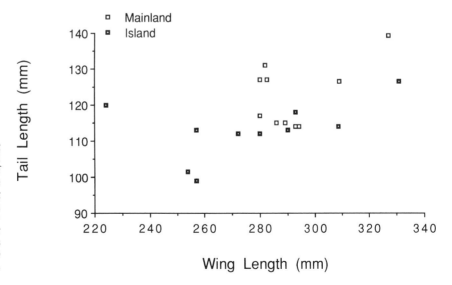

Fig. 2.10. Comparison of wing and tail lengths of mainland and small island subspecies of barn owls. Island forms are significantly smaller: for wing length, $z = -1.72$, $p = 0.04$; for tail length, $z = -2.88$, $p = 0.002$. Based on smaller sample sizes, tarsus lengths are also less; $z = 1.77$, $p < 0.04$ (see Appendix 1).

wings ($p = 0.03$), tails ($p < 0.01$) and tarsi ($p = 0.05$), and, although there is little information on body weights, there can be no doubt that they are substantially smaller than mainland forms. The smallest seems to be the Galapagos Islands subspecies which has a wing length of only 224 mm, about two-thirds that of the North American birds and a body weight of only around 260 g. The forms on Sao Thome in the Gulf of Guinea, on the Andaman Islands and on the Canary Islands are also exceptionally small. Island subspecies often have a narrower range of prey available to them with insects frequently becoming important components of their diet (Chapter 4) and it is possible that natural selection has therefore favoured small body size in these circumstances.

General body coloration above and below varies greatly, the nominate form, as its name suggests, being among the palest. Generally, paler plumage seems to have arisen among forms that inhabit more open savannah or grassland environments. Extremely dark plumage has developed mainly in two circumstances: among subspecies that live in more forested habitats such as the south-east Asian form, *javanica*, and among those inhabiting small islands. Many of these islands are tropical or sub-tropical and must originally have been mainly forested before the recent effects of humans so it is difficult to be sure if dark plumage is associated with islands or forest. Examples of dark island forms are those of Sao Thome, the Andaman Islands and the Lesser Antilles Islands all of which probably utilise more forested environments than the nominate form. Dark plumage would clearly be an advantage as hunting camouflage or as protection against predators in the lower light levels among trees. Paler plumage, especially of the underparts, might be adaptive in a similar way to the pale underparts of many seabirds by giving less contrast against the sky when hunting. This might not seem useful to a largely nocturnal predator but it could be important on bright moonlit nights. Some of the dark island forms such as the small Galapagos barn owl clearly do not hunt in forested habitats and their plumage coloration may have evolved completely independently to match the dark volcanic rocks amongst which they live. Here again dark plumage would give good camouflage. The central European *guttata* subspecies is also darker than the nominate form (Fig. 2.11), but the significance of this difference is unclear.

2.3 Related species in the genus *Tyto*

The greatest radiation within the genus *Tyto* has taken place in the Australasian region where at present seven or eight species occur including *Tyto alba*. The group has diverged considerably in morphology,

Fig. 2.11. The *guttata* subspecies of central and eastern Europe has much darker plumage than the nominate *alba* subspecies. (Photograph: Johan de Jong.)

behaviour and ecology to occupy a range of habitats from dense tropical grasslands to rainforests (Schodde and Mason 1980, Hollands 1991). None of these other species is as well-known as *T. alba* and for some, especially those living in rainforest habitats with restricted island distributions, we have almost no ecological information.

Closest to *T. alba* in general anatomy and body weight is the grass owl which has a highly discontinuous distribution in Australia, southeast New Guinea, Sulawesi, the Philippines, Taiwan, southern China, India and Pakistan, and Africa. The isolated African form *T. capensis* is sometimes considered to be a separate species from the Australasian forms, together grouped into *T. longimembris*. Adapted to living in tropical and sub-tropical grasslands in a way similar to the short-eared owl of northern areas, the grass owl has exceptionally long wings and legs, hunts its rodent prey in flight by slow quartering and nests on the ground. A flimsy nest of stems may be built which is reached by means of tunnels through the adjacent grasses. Clutch size may vary from three to eight and breeding success and numbers seem to be related to rodent abundance.

The masked owl, *T. novaehollandiae*, of Australia and southern New Guinea is highly variable in size and colour but is generally a very large owl with body weights ranging from 520 to 1260 g. Tasmanian birds seem to be the largest. In marked contrast to *T. alba* and the grass owls, there is considerable sexual dimorphism in body size with females much larger than males. It inhabits open woodland including some wooded farmland and nests in large tree cavities. Its diet is highly variable including terrestrial marsupials, such as various *Antechinus* species and rodents, and arboreal species, such as ring-tailed possums (*Pseudochirus peregrinus*). Clutch size normally ranges from two to four.

Perhaps the most dramatic of all the *Tyto* species is the sooty owl (*T. tenebricosa*), an inhabitant of moist eucalypt forests of south-east Australia and montane rainforests of New Guinea. It is a large species with body weights ranging from 500 to 700 g in males and from 750 to 1000 g in females. It has sooty black upperparts and either sooty or grey underparts, all finely flecked in white. The eyes are very large and dark and the talons are massive. Superbly adapted to operate in the darkness of the forest environment, it catches mostly arboreal marsupials some of which, such as the greater glider (*Petauroides volans*) and ring-tailed possum, are very large and must need considerable power to kill. The nest is usually in a hollow tree, although sometimes cliffs or caves are used. Normally only one egg is laid but sometimes two.

The lesser sooty owl (*T. multipunctata*), found only in the rainforest remnants of Queensland, is often considered to be a subspecies of the sooty owl but its morphology and ecology are sufficiently different to merit full species status. It is smaller than the sooty owl, at around 450 to 550 g, and hunts much smaller prey, mostly mammals but also some insects and birds, taken mainly from the forest floor. Large tree cavities are used for nesting and usually two eggs are laid.

The remaining members of the genus include the Celebes, or Sulewesi barn owl (*T. rosenbergii*), the Minahassa barn owl (*T. inexpecta*) which is found only in the northern peninsula of the same island, the New Britain barn owl (*T. aurantia*) and the Malagasi grass owl (*T. soumagnei*).

2.4 Summary

In total, 36 subspecies of the barn owl have been recognised. They vary considerably in size, body proportions and coloration. Island forms tend to be smaller than mainland forms and among the latter the south-east Asian subspecies is by far the largest. The North American form has a wing length about 12% greater and tarsus length almost 25% greater than the European subspecies. Subspecies inhabiting more forested regions tend to have shorter wings and darker plumage whereas those that live in more open savannah and grasslands have longer wings and lighter plumage. Subspecies of some volcanic islands such as the Galapagos Islands also tend to have dark plumage. Leg length tends to be greater in forms that forage over longer tropical and sub-tropical grasslands than in those that hunt over shorter, temperate region grasslands.

Barn owls show little size dimorphism between the sexes; males either do not differ from females in wing, tail or tarsus length as is the case in Europe, or show extremely small differences of a few millimetres that are only statistically significant when large samples are analysed. However, females tend to be heavier than males, especially during the breeding season. Generally, females are darker above and below than males and have more pronounced flecking on their underparts and stronger wing markings. Females tend to become paler with age.

There are seven (or perhaps eight) other members of the genus *Tyto*, most of which are found in the Australasian region where they inhabit both grassland and dense forest habitats. Some of these, such as the masked owl, show considerable size dimorphism between the sexes with females larger than males.

3 *Diet*

The life history patterns of all predators, their growth and development, reproduction, longevity and movements are profoundly influenced by the food they eat and how and where they obtain it. Depending upon its size and anatomical and behavioural adaptations, each species catches a particular range or spectrum of prey, but within that range there are usually many variations that are of great interest as they often provide telling insights into the ecological significance of diet. The barn owl diet has been studied in more detail and more extensively than that of any other bird of prey, a reflection of the wide distribution of the species and the ease with which pellets containing the regurgitated, indigestible remains of prey items can be found and analysed. Such pellets can be collected in large numbers at nesting places during the breeding season and at roost sites at other times, enabling observers to make comparisons of different areas and seasons (Fig. 3.1).

The pH of the barn owl's stomach is higher (less acidic) than that of many other predatory birds with the result that most of the bones of ingested prey are left undigested (Smith and Richmond 1972). The skulls and mandibles of even the most delicate small mammal and bird prey are found intact in the pellets and can easily be identified. The proportions of various prey recovered in this way are usually in very good agreement with the proportions consumed. Trials comparing known diets fed to captive birds with the remains subsequently found in their pellets have shown close similarity (Raczynski and Ruprecht 1974, Dobson and Wexlar 1979). In my Scottish study area, I could detect no significant difference between the prey items found at the nest and the contents of pellets (Fig. 3.2). In the same area, continuous photographic records were made by Ian Langford of all prey delivered to four nests over several nights, which were then compared with the items recovered from the pellets produced over the same period. Again there was very close agreement. Some researchers have reported decapi-

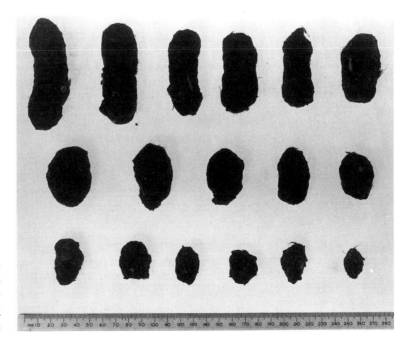

Fig. 3.1. Barn owl pellets vary greatly in dimensions. Analysis of their contents can give reliable information on diet. (Photograph: Iain R. Taylor and Dave Haswell.)

Fig. 3.2. Prey items stored at the nest can give information on diet. This collection consists of field voles (top row), common shrews and a swallow. (Photograph: Iain R. Taylor.)

tated prey being delivered to nests (e.g. Pikula *et al*. 1984) and this has been recorded in photographic sequences at a number of nests in Switzerland (Michel Juillard, personal communication) but none of 614 prey deliveries photographed in Scotland involved decapitated items. Just how frequent or widespread decapitation might be is unknown, as is its function, but it could introduce an element of error in some areas if pellet analysis was based on skull and jawbone identification alone.

Diet studies have been conducted in almost every continent and in many different habitat types. Their number runs into many hundreds, all of which contain interesting and useful information but, regrettably, space does not permit a comprehensive and exhaustive exploration of all of them. Instead, I have searched through the literature to try to discover studies that give a representative picture of the barn owl diet in as many major regions as possible. I have had to be highly selective and have given preference to studies that have involved substantial numbers of birds over a reasonable number of years in an attempt to reduce biases that undoubtedly occur in less extensive research programmes. For some areas, there have been no long-term studies and rather than leave large gaps I have used shorter-term studies provided they involved a reasonably wide sample of the population.

In this way, I have identified 52 studies from most parts of the barn owl's range and these form the basis of much of the subsequent discussion. The results of many were originally presented only as the numbers of each prey species consumed. In these cases I have reanalysed the data, using prey weight information from similar studies, to derive a quantification of diet in terms of biomass in addition to numbers, and to make estimates of the means and ranges of prey weights. All of these data are presented in Appendix 2. The results of other studies have been referred to only when they provide additional information on a particularly interesting or important point.

3.1 Diversity of prey in the diet

With very few exceptions, most studies have shown barn owls to prey mainly on small mammals. In the 52 sample studies that I analysed, the contribution of these prey ranged from 74 to 100% of all items caught and in 42 (79%) cases they made up more than 90% of the items taken (Fig. 3.3). Barn owls inhabiting relatively unproductive, hot, dry parts of the world tend to rely slightly less on small mammals and those from moist temperate regions tend to specialise very highly upon them, but this is by no means a hard and fast rule. For example, diets consisting almost entirely of small mammals have been recorded in some semi-arid parts of Australia, but even here this might not be a real exception as the analyses were made mostly when individual rodent species had

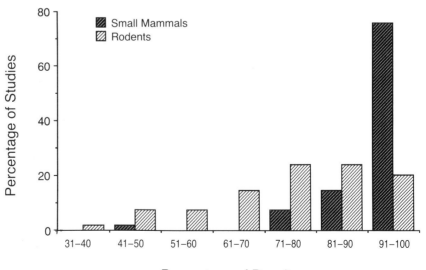

Fig. 3.3. Proportion of small mammals and rodents in barn owl diets around the world (data from 52 studies detailed in Appendix 2).

reached high or 'plague' population levels, thus attracting the researchers' attention to the areas. Under more normal conditions the birds might well resort to a wider variety of prey types.

The great majority of small mammals taken are terrestrial species but the capturing of bats is widespread involving a great diversity of species, although bats never contribute significantly to the overall food intake of the owls (Ruprecht 1979a, b, Bersuder and Kayser 1988). Rodents are by far the most important small mammal group: in 47 of the 52 key studies they formed at least half of all prey items and in 33 (63.5%) they made up more than 75% of all prey (Fig. 3.3).

The usual alternatives to small mammal prey are birds, lizards, amphibians and insects. Insects in particular seem to feature significantly in hotter, drier environments. Interestingly, in temperate regions, barn owls do not catch earthworms, even when they are abundant and other prey are scarce.

Extreme exceptions to the normal diet range are uncommon and usually occur in situations where small mammals are absent or scarce or where an alternative prey source is very abundant. Thus, on the Cape Verde Islands barn owls feed on reptiles, mostly geckos, and a wide variety of birds including plovers, godwits, turnstones, pratincoles and weavers (de Naurois 1982). On a rocky island off the Californian coast, Bonnet (1932) discovered barn owls nesting in an old cabin. Four young were reared, apparently mainly on a diet of the locally breeding Leach's storm petrel (*Oceanodroma leucorhoa*), the remains of which formed a carpet 7.5 cm deep on the cabin floor. Other studies have shown individual specialist bird-eating barn owls, usually where their hunting ranges include the large communal roosts of species such as red winged blackbirds, *Agelaius phoeniceus*, in North America (Carpenter and Fall 1967).

Barn owls will also take advantage of other temporarily abundant prey types that are very different from the normal diet. In West Africa, I often observed groups of them running around on tarmac roads catching winged termites that had been attracted to the road lighting and had accumulated in vast numbers, intent on their sexual chases. Although small, termites are of very high nutritional value and the owls could catch large numbers quickly with little effort. There is also an astonishing record of a group of 30 barn owls catching a species of fish, the grunion (*Leuresthes tenuis*), which had gathered in great numbers on a beach in southern California (Bent 1938).

The total number of small mammal species taken in most diets, excluding those species contributing less than 0.5% of items, ranges from 2 to 25 in different parts of the world indicating that the diversity of the barn owl diet, or its 'foraging niche width', varies greatly from place to place. The breadth of a species' diet could have a profound effect on many aspects of its ecology and is therefore of considerable interest but, before it can be investigated, an objective method of measuring it is needed. Hererra (1974) and Marti (1988) did this by applying mathematical formulae that took account of both the number of species in the diet and their relative abundance, resulting in a single numerical value for diet diversity. There are many advantages in this approach but there are also disadvantages. A simple numerical value might enable comparisons to be made but on the other hand a great deal of useful information is lost and there is also sometimes a risk of implying a greater level of precision than is justified by the potential biases in the original data. In order to appreciate the ecological significance of a single figure index, it is also necessary to return to the original data. Only then can the real nature of the diet be seen.

Most barn owl diets tend to be dominated by a relatively small number of species and to have a long tail of species that are taken regularly or sporadically but which, in total, contribute only small amounts (Appendix 3). The length of this tail also depends on the number of prey items contained in the analysis (Baudvin 1986). Examination of the 52 sample studies indicated that approximately the last 10 to 20% of most diets was made up of many species of which none contributed more than a few per cent (Fig. 3.4). Thus, it seemed quite meaningful to make comparisons of diets based on the minimum number of species needed to form at least 80% of all items. This can be expressed as numbers or as biomass and using this method it can be seen whether a population of owls depends mainly upon one, two, three or whatever number of prey species. Major differences in diet diversity among populations or

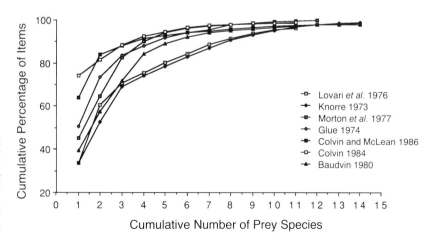

Fig. 3.4. In most barn owl diets a relatively small number of species make up about 80% of items, the remaining long 'tail' being composed of many less numerous items. A representative sample of diets is shown.

subspecies should be revealed by this approach. A comparison of values obtained in this way with the indices given by Marti (1988), derived from the more complex calculation, for 12 studies showed that they were almost exactly correlated (Spearman rank correlation, $R_s = 0.96$, $p < 0.001$). Using the simpler method is therefore just as valid for comparative purposes and it is somewhat easier to visualise its ecological meaning. I have used these methods in the following analysis that has been subdivided geographically to show precisely which species are important in different parts of the barn owl's range. Full details of a representative sample of diets from different areas is given in Appendix 3.

Barn owl diets in Europe

Throughout Europe, three major groups of small mammals, voles (Microtidae), shrews (Soricidae), and mice and rats (Muridae) predominate in barn owl diets but their relative importance varies from region to region (Fig. 3.5, Appendix 2). In Britain and most of central continental Europe, voles are by far the most important prey both in numbers and biomass. The field vole (*Microtus agrestis*) is the dominant species on mainland Britain, on average contributing about 50 to 65% of the weight of prey eaten. On the continent, field voles are still taken but almost always in low proportions, and the common vole (*Microtus arvalis*) takes over as the most important species. Common voles are absent from Britain except for the Orkney islands and Guernsey where they were probably introduced by early human colonists.

These two voles differ markedly in their habitat selection, the field vole preferring moist long grassland whereas the common vole frequently occurs in drier, shorter grasslands such as pastures and also in crops such as cereals. This difference has important implications for the owls' foraging behaviour and for the conservation of the species and is discussed more fully in following chapters. Some voles, such as the bank vole (*Clethrionomys glareolus*) and the common pine vole (*Pitymys subterraneus*) although common in the general environment in which the owls hunt, are usually taken infrequently, presumably because their behaviour makes them relatively unavailable to the owls. The bank vole prefers to remain within dense vegetation such as woodland and hedgerows and the pine vole spends a great deal of its time underground.

Shrews are nearly always the second most frequently taken prey in Europe. In Britain these are entirely the red-toothed shrews of the genus *Sorex*: mainly the common shrew (*Sorex araneus*), although the diminutive pygmy shrew (*Sorex minutus*) and larger water shrew (*Neomys*

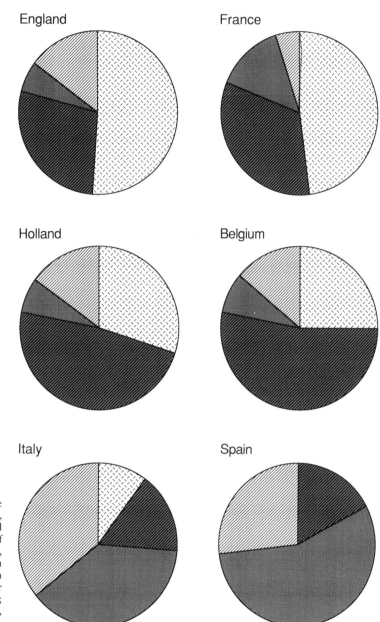

Fig. 3.5. Variation in the relative importance of voles, shrews and mice in barn owl diets in different parts of Europe. Sources: England, Glue 1974; France, Baudvin 1983; Holland, de Bruijn 1979; Belgium, van der Straeten and Asselberg 1973; Spain, Herrera 1974; Italy, Lovari *et al.* 1976.

fodiens) are also taken occasionally. The last is hunted where streams flow through the bird's foraging range. On the continent, common shrews are also important prey but members of the white-toothed shrew genus, *Crocidura*, which are absent from Britain, also become significant, especially the common white-toothed shrew (*Crocidura russula*). In Holland, Belgium and Denmark, shrews are often the most frequently taken prey, with the common shrew making up about 30–40% of items. In these areas, the common vole is usually ranked second by numbers but because of its greater weight contributes about equally with the common shrew to total prey biomass.

To the south, in Mediterannean climates, where conditions are generally unfavourable for the growth of succulent grasses, common voles are absent and field voles are greatly restricted in their distribution. Here, barn owls depend mainly upon various species of mice, particularly the house mouse (*Mus musculus*), African house mouse (*M. spretus*), wood mouse (*Apodemus sylvaticus*) and yellow-necked mouse (*A. flavicollis*). In the driest areas of Spain, southern France, Crete and Africa, species in the genus *Mus* predominate whereas in the somewhat cooler and less arid parts, such as much of Italy, *Apodemus* species are more important. Amongst the shrews, various *Crocidura* species are often important secondary prey in the drier areas and *Sorex* species in the more moist regions.

This relationship between climate, vegetation types and barn owl prey species is clearly demonstrated by the results of a study in the south of France around Perpignan in which the owl diet was quantified at a large number of sites ranging from sea level to about 1400 m (Libois *et al.* 1983). At low levels the vegetation was mainly xerophytic, Mediterranean dwarf shrub and herb communities, but with increasing altitude it became progressively less dry with the increasing prevalence of grassland and deciduous woodland. The barn owl diet showed considerable variation according to local conditions and was slightly complicated by irrigation areas at low altitude but nevertheless trends in relation to climate and vegetation were clear. Various species of mice in the genus *Mus* formed about 10 to 25% of the items taken in the low altitude dry areas but their contribution decreased with altitude and eventually they disappeared entirely at sites above about 700 m. At low altitude *Apodemus* species generally contributed less than 30% of the diet and this increased up to 60% in the less dry upland sites. *Sorex* species were completely absent at low altitudes but made up a significant proportion in most upland areas and the field vole was also more important in the moister upland regions.

In many Mediterranean areas, bank voles and pine voles are captured more frequently than in most temperate areas and are often significant components of the diet. It is not known whether this results from a change in the habitat preferences or behaviour of these voles in drier areas or from a change in the foraging selection of the owls. It is possible that with reduced competition from field voles, bank voles may move into more open habitats.

Mice are important prey in two other regions of Europe: Ireland and the Isle of Man, and some parts of eastern Europe. Field voles and common shrews are absent from Ireland and the Isle of Man and here wood mice and house mice are the most important prey. Bank voles were first discovered in south-west Ireland in 1964 (Classens and O'Gorman 1965) and in this area they are now captured frequently by barn owls (Smal 1987). Again it is possible that, in the absence of other voles, bank voles may have been able to spread into habitats where they are more vulnerable to the owls.

Several studies in eastern Europe have shown house mice to be the dominant prey species and house sparrows (*Passer domesticus*) are also taken frequently in these areas. The regions concerned are in farmland where traditional methods of agriculture are still practised and where there is little use of more recently developed practices of harvesting and food storage. As a result these commensal species are more abundant than in most other parts of temperate Europe.

Considering the whole of Europe, the number of prey species in the barn owl diet (excluding those contributing less than 0.5%) varies from 5 to 18. Values are lowest in Britain and Ireland with from 5 to 11 species, highest with up to 18 species in central Europe and again somewhat lower in Mediterranean areas with about 10 to 15 species. This is partly a reflection of the different number of species available to be caught: central Europe is particularly rich in small mammals. The number of species making up 80% of the diet (by number of items) varies from one to six with three or four being the most frequent (Fig. 3.6). There was a slight but significant negative correlation between diet diversity, measured in this way, and latitude ($R_s = -0.343$, $n = 24$, $p = 0.05$). Thus, more southerly barn owl populations in drier and less productive habitats tend to have a slightly higher diversity of prey in their diet (Fig. 3.7). A similar tendency was noted in an earlier analysis by Herrera (1974). The situation is, however, complicated by altitude and local habitat effects so there are many variations. When expressed by biomass instead of numbers, the owls were mainly dependent (80% of biomass) on from two to five prey species in 23 of the 26 studies examined in Appendix 2 and in 16 of these they were dependent on only three or four species. Aside from the latitude effect, diets seem to be much less diverse where they are dominated by voles than where other species are more important. The mean number of species making up 80% of all items (by number) was 3.4 in vole-dominated diets but 4.5 in non-vole-dominated diets (Mann-Whitney U-test, $z = -2.22$, p

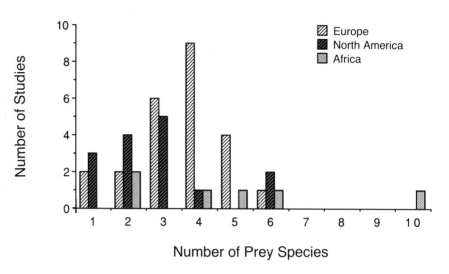

Fig. 3.6. Comparisons of diet diversity in Europe, North America and Africa. The graph shows the number of species making up 80% of all items. Diets tend to be most diverse in Africa and most specialist in North America. Data from Appendix 2.

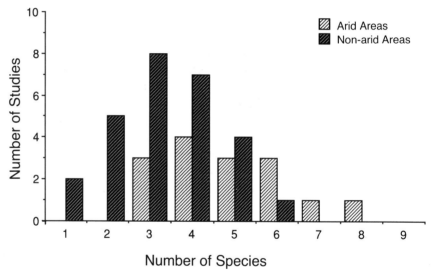

Fig. 3.7. Comparison of diet diversity (number of species contributing 80% of items) in arid or semi-arid environments with non-arid regions worldwide. The barn owl diet tends to be narrower in more productive habitats.

= 0.013). Where either the common vole or the field vole was the most frequent prey, the single species made up between 40% and 90% of the total biomass of prey consumed. In diets where voles were not the most frequent items, no single species ever contributed more than 40% of the biomass.

Diets in North America

Fifteen studies were included in the analysis for North America, representing a wide range of regions and habitats (Appendix 2). As might be expected for such a large continent, the species content of the owl diet varies greatly from state to state. Depending upon the region, the important species include voles (*Microtus* spp.), shrews (*Sorex, Blarina* and *Cryptotis* spp.), pocket mice (*Perognathus* spp.), white-footed mice (*Peromyscus* spp.), harvest mice (*Reithrodontomys* spp.), pocket gophers (*Thomomys* and *Geomys* spp.), kangaroo rats (*Dipodomys* spp.), wood rats (*Neotoma* spp.), rice rats (*Oryzomys* spp.) and Old World rats and mice (Muridae).

In most northern states, down to about latitude 35° N, voles are by far the most important prey both in numbers and in biomass. East of the Mississippi, the meadow vole (*M. pennsylvanicus*) is the main species and further west in the Rocky Mountain areas of Colorado, Idaho and Utah the mountain vole (*M. montanus*) and the prairie vole (*M. ochrogaster*) also become important. Further west still, on the Pacific coast, in states from Oregon to British Colombia, the Oregon vole (*M. oregoni*) and Townsend's vole (*M. townsendi*) are the main species. These changes simply reflect the distribution patterns of the different vole species.

Compared with European barn owls, the North American birds seem to depend much more on single species of prey (Fig. 3.6). Thus, in vole-dominated diets, the average contribution of the main vole species to the total diet biomass was 71.2% ($n = 9$) in North America compared with 54.5% ($n = 15$) in Europe (Mann-Whitney U-test, $z = 1.94$, $p = 0.026$). In these mainly vole diets few other species are of major significance although many species contribute to some extent. In Eastern areas, house mice, rice rats, short-tailed shrews and white-footed mice are the usual secondary species. In the Rockies states, white-footed mice, pocket mice, harvest mice, pocket gophers and house mice are important and along the Pacific coast various shrews of the genus *Sorex* are often taken.

Most voles are absent from the southern states and so other small mammals form the owls' diet. In the south-east, from Texas across to South Carolina, the cotton rat (*Sigmodon hispidus*) is usually the most frequent prey but most diets are more diverse than those further north with up to seven or eight species making up 80% of prey items. Thus harvest mice, pocket mice, white-footed mice and the least shrew are normally all important components of the diet. In Oklahoma, in mesquite habitat, the southern plains wood rat (*Neotoma micropus*) and the plains pocket gopher (*Geomys bursarius*) are important in addition to the above species. The increasing diversity of the owl diet with decreasing latitude is similar to that observed in Europe.

Diets in Australia

The diet of Australian barn owls has been studied in the Northern Territory, south-west Queensland, South Australia, Victoria and on islands off the coast of Western Australia; (Appendix 2; Valente 1981, Dickman, Predavec and Lynam 1991). In the last three of these areas, the owls depended almost entirely on house mice with small proportions of a range of indigenous rodents, especially the plague rat (*Rattus villos-issimus*), the hopping mouse (*Notomys alexis*) and dasyurid marsupials. Two of the studies, in Queensland and Northern Territory, recorded plague rats and the hopping mouse as the dominant prey. Dasyurid marsupials never seem to feature significantly: Morton (1975) and Morton *et al.* (1977) recorded a rodent to dasyurid ratio of about 1 to 20. This is probably a consequence of differences in the abundance of these groups. Australian barn owls, therefore, like those in Europe and North America seem to be highly dependent upon just a few species of rodents. In all studies, a single rodent species made up more than 50% of the diet biomass. It must be stressed, however, that most of these studies were done when the populations of these rodents were at relatively high abundances. More long-term work is needed to discover how the barn owl diet changes when these become less numerous.

Diets in Africa

African barn owls depend heavily on rodents, which usually form between 80 and 90% of all items consumed. Compared with other areas however, their diets tend to be more diverse. In the Transvaal, for example, up to 25 small mammal species are regularly caught. Frequently, five or six species make up 80% of all items taken compared with the normal three or four in Europe or two or three in North America (Fig. 3.6). Diets are particularly diverse in the more arid areas in the Kalahari desert of south-west Africa and Cape Province as many as 10 species may be involved in forming 80% of the diet biomass.

Throughout much of the non-arid regions, the multimammate rat (*Praomys natalensis*) is a dominant or important species, often forming up to 60% of the total prey weight. The vlei rat (*Otomys irroratus*) and other members of the genus are taken widely and sometimes are the main prey species. In arid areas, members of the gerbille genera *Tatera*, *Gerbillus* and *Desmodillus* species are usually the dominant prey. House mice and black rats can be important species in some areas such as Morroco and Egypt (Saint-Girons and Thouy 1978, Goodman 1986).

Shrews are usually the most numerous non-rodent prey in all except the arid regions, but only in one study in Tanzania (Laurie 1971) were they the dominant prey. In Natal and Transvaal they formed 9% and 7% of the items caught.

Birds feature widely in barn owl diets throughout Africa. Mostly these are weaver birds (Ploceidae) which roost in large numbers either at their nesting places or in reed beds and hence become vulnerable to owl predation.

Lizards, as a consequence of their largely diurnal habits, are rarely caught by barn owls but they are sometimes taken in arid areas. At one site in the Namib desert they formed just under half of the items taken (Vernon 1972).

Diets in other areas

Information on the barn owl diet in other parts of the world tends to be more sketchy, and for many areas I have been unable to find any details. Despite its wide distribution throughout much of south-east Asia, studies seem only to have been undertaken in Malaysia. Here, the barn owl is found in oil palm plantations and in coastal rice growing areas and throughout the area feeds mainly on indigenous *Rattus* species such as the Malaysian wood rat (*Rattus tiomanicus*). This species which is a serious pest of palm fruits can form up to 90% of all items consumed and for this reason the owls' presence is greatly encouraged by the provision of nest boxes (Duckett 1976, Lenton 1980, Smal, Halim and Amiruddin 1990).

Rats and mice are often important prey for island populations of barn owls. On the Galapagos Islands, the black rat (*Rattus rattus*) made up about 34% of the biomass of prey with the house mouse contributing about 48% (de Groot 1983). Black rats formed 77% of the items recovered from pellets collected on five islands in the southern Bahamas (Buden 1974) and were the most numerous prey on Grand Cayman in the West Indies (Johnson 1974). On the Canary Islands, black rats and house mice together accounted for 93% of the biomass of prey (Martin, Emmerson and Ascanio 1985).

The barn owl's ability as a rat catcher has resulted in it being introduced onto several island groups to control these rodents but things have rarely turned out as planned. The North American subspecies was established on Hawaii betwen 1959 and 1963 (Berger 1972), but instead of specialising on rats it fed mostly on house mice (Tomich 1971). The African subspecies was introduced into the Seychelles during the 1950s to control rats in palm plantations but quickly took to catching rarer native small mammals and protected birds such as the fairy tern (*Sterna nereis*) resulting eventually in efforts to eradicate the owls.

On many islands, especially the drier ones, barn owls often catch a greater proportion of non-mammal prey, usually birds and insects, than in mainland areas. On the Galapagos Islands, about half of the items taken were insects, although because of their small size compared with rodents they only contributed 12% of the diet biomass. On the Canary Islands, insects made up 18% of the items but only just over 1% of biomass. This increased use of insects is probably the result of a shortage of other larger items and the smaller size of island subspecies of the barn owl (Chapter 2) may be an evolutionary response to a dependence upon these small prey. Presumably there must be times when the owls depend on them even more than was shown in these studies and during such periods natural selection would be expected to operate against individuals of large body size.

3.2 Diet in relation to habitat

Traditional farming and forestry practices gave rise to highly varied landscapes over much of Europe and North America and, although modern methods have tended to generate uniformity, it is still possible to find regions where local diversity remains. Barn owls in these areas live in a great variety of conditions and individuals even only a few kilometres apart can be in completely different habitat types. Thus, in my Scottish areas, pairs nesting in pastoral farmland with intensively cultivated fields, hedgerows and small woods had neighbours close at hand living in the uniform habitat of young conifer plantations. The nature of the habitat can influence the relative proportions of different species in the local small mammal community and this in turn can be reflected in the owl diet. A population of owls in such a diverse area is in effect living in a fine mosaic, with patches of greatly differing quality being occupied by the individual pairs.

In Scotland, the range of mammal prey available to the owls was not great, but nevertheless there were significant relationships between habitat and diet. In the upland conifer plantations, only two main prey species were available, field voles and common shrews, and these were the only significant items in the barn owl diet. Probably as a result of the uniformity of this habitat most pairs took very similar proportions of these prey. The rough grassland habitat of the sheepwalk areas again had only field voles and common shrews available in significant numbers, but in this case there was considerable variation in diet among the owl pairs. Those whose range enclosed a high percentage of moderate to long vegetation had a dominance of field voles in their food, whereas pairs living in areas with more closely cropped and tussocky grassland had a much higher proportion of common shrews. The lower altitude pastoral farmland area with its numerous small woods offered the owls an additional species, the wood mouse, and this became a significant item in the diet, especially when voles were temporarily scarce. Therefore, even though the owl diets were all very restricted, there were small differences between habitats. These were important to the owls and had a significant influence on productivity and population dynamics, as will be shown in subsequent chapters.

Most studies that have covered large enough areas have demonstrated links between local habitat variation and the owl diets often more complex than those described above. In Holland, for example, the owls had a restricted diet consisting mainly of common voles and common shrews (together making up from 80% to 100% of items) in the so-called 'vole plague areas', extensive plains of lightly grazed pastures and fields in polder or riverine areas. In more elevated parts with more varied landscapes, they regularly took four or five species including common vole, common shrew, white-toothed shrew, wood mouse and field vole (de Bruijn 1979).

A very similar variation has been described in Norfolk, England where field voles and common shrews were the only significant prey in pastoral areas but other species such as wood mouse, brown rat and bank vole became more important in the more varied landscapes of mixed farmland with hedges and woodlands (Buckley and Goldsmith 1975).

3.3 Seasonal changes in diet

Most barn owl populations experience some seasonal change in their diets, the few apparent exceptions being those in some tropical environments such as in Malaysia where climatic conditions tend to be more stable throughout the year.

In Scotland, the proportion of field voles in the barn owl diet was lowest in the spring months of March to May and increased during the summer to reach a peak in late autumn and early winter, more or less following the annual change of adult vole numbers in the wild population. As might be expected, this pattern has been recorded many times and seems to be a widespread phenomenon in temperate regions (Webster 1973, Dawe, Runyan and McKelvey 1978, Baudvin 1983, de Jong 1983, Campbell, Manuwal and Harestad 1987) although some studies, often of shorter duration, have found no clear change with season (e.g. Smith, Wilson and Frost 1972).

The main alternative to voles in much of Europe is usually shrews of various species and often these feature more in the owl diet in winter and spring than during summer. The difference is particularly marked in Holland where in one area studied by de Bruijn (1979) common shrews increased from 28 to 40% of the diet biomass from summer to winter and white-toothed shrews increased from 5 to 12.5%, while the contribution of common voles fell from 48% to 30%.

These changes in diet mean that voles, which are generally the heaviest prey items and probably the most profitable (Chapter 6), feature least in the owl diet just exactly at the time of year they are starting to breed and when their energy and protein demands are high. However, the voles caught at this time are mostly very large individuals (Chapter 6, Brown 1981) so this might partly offset any disadvantage from their relatively low frequency in the diet.

In North America, a great variety of seasonal changes have been recorded in addition to those described for voles. Species such as pocket gophers and pocket mice can be important prey during summer months but as the former retreat to deep burrows in winter and the latter often hibernate they are seldom taken in winter (e.g. Marti 1974). Behavioural differences between species can alter their relative susceptibility to predation at different seasons. Thus, in coastal saltmarsh habitats of New Jersey, rice rats which form only 2% of the summer diet increase to make up 23% of the winter food, whereas meadow voles decline from 91% in summer to 58% in winter. The change seems to arise from different behavioural responses to snow cover. Voles are mainly active under the snow and hence their availability to the owls is reduced, whereas rice rats remain active on the snow surface and their vulnerability to predation is increased (Colvin 1984).

Dramatic seasonal changes can also occur in response to variations in rainfall. A population of owls studied over 4 years in Sicily was mainly dependent upon small mammals but each year for about 2 months, coinciding with the hot dry period from about July to September, the owls greatly increased their predation on insects which consistently formed 70–80% of the items caught. The reasons for this may have been quite complex, involving an increased abundance of large insects and reduced availability of small mammals but also, at that time of year, the owls had completed breeding and were not constrained by the need to carry larger items such as small mammals back to the nest (Migliore, Rizzo and Massa, University of Palermo, Italy, unpublished).

3.4 Long-term changes in diet

I quantified the diet of most pairs in the Scottish study population by pellet analysis over 12 consecutive years which encompassed four complete vole cycles. Pairs living in the conifer plantation and sheepwalk habitats were almost totally dependent upon only two prey, field voles and common shrews, whereas for those nesting in low farmland habitat wood mice were also important. In each habitat, the relative contributions of these prey species to the diet varied greatly from year-to-year.

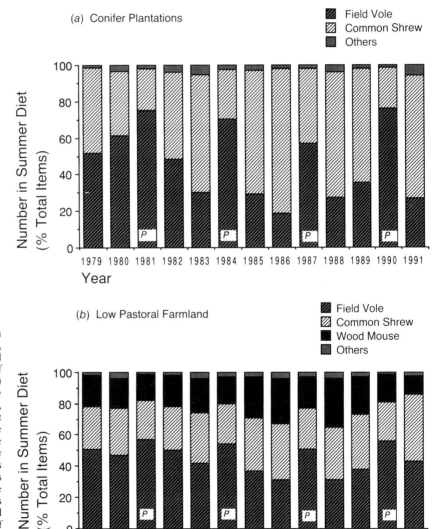

Fig. 3.8. Annual variations in the percentage of field voles, common shrews and wood mice in the summer diets of barn owl pairs nesting in (a) conifer plantation and (b) low farmland, Esk study area, south Scotland. Highest proportions of voles were taken in years of peak vole abundance (P) and a distinctly cyclic pattern is evident. Pairs nesting in sheepwalk habitats had diets similar to those nesting in plantations. Diets were least diverse in the good field vole habitats of plantations and more diverse in the less good low farmland habitat.

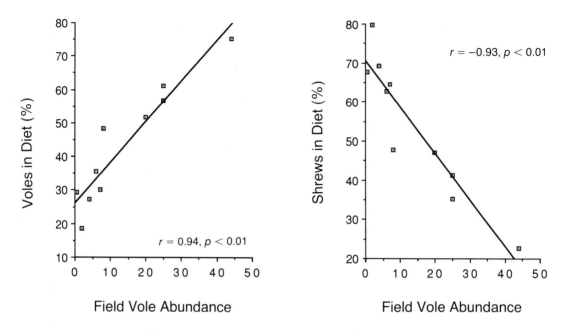

Fig. 3.9. Correlation between the percentages of field voles and common shrews in the summer diets of conifer plantation nesting pairs each year and the abundance of field voles.

The variations were cyclic and correlated strongly with the cyclic variations in the abundance of field voles (Figs. 3.8, 3.9). In plantations, at peak vole densities, voles made up about 75% of all items and 90% of the biomass of the owl diet. As vole abundance declined, the owls captured increasingly greater numbers of shrews so that, at the lowest vole densities, shrews made up 45 to 50% of the biomass consumed compared with only about 10% at vole peaks. Almost identical changes occurred in the sheepwalk habitat. In low farmland the proportion of voles caught also varied cyclically but the amplitude was much less pronounced than in the other habitats. In addition to switching to common shrews the birds also caught increasing numbers of wood mice.

Similar cyclic changes in the importance of voles have been recorded in a number of other studies. In Germany, Bohnsack (1966) found that common voles made up almost 95% of items in peak vole years but only 26% in low vole years. Again, the common shrew was the most important alternative prey but several other species were also taken more frequently when voles declined. In Holland, de Bruijn (1979) compared diet changes in areas prone to large fluctuations in vole abundance, 'vole plague areas', with those in areas where less extreme vole cycles occurred. In the former, common voles varied from 42 to 90% of items in the diet over the vole cycle and in the latter, from about 15 to 50%. Common shrews were the main alternative in 'plague' areas but in the other areas the common white-toothed shrew was much more important. Similar cycles in diet have also been recorded in Poland (Goszcynski 1981) and are probably of widespread occurrence over much or possibly even all of Europe where voles predominate in the diet.

Annual diet variations of this type are very much less pronounced in the barn owl populations that have been studied in North America. Studies over 8 years in Utah showed that microtine voles made up from 70 to 85% of the diet in 7 years and 55% in the remaining year with no evidence of a cyclic pattern. More variation was evident in an Idaho population, from 36–65% over 8 years, but again no cyclicity was obvious (Marti 1988). The vole populations in these areas, in marked contrast with the European areas, were not cyclic (Carl Marti, personal communication). The habitats involved were, however, somewhat unusual being mostly irrigated farmland in otherwise dry desert environments and it would be interesting to have results from more temperate habitats.

It is rarely possible to examine changes in the barn owl diet over historic timescales, but a series of studies enables this to be done for England. In the 1920s and 1930s, brown rats were important prey making up about 38% of the diet biomass in two studies (Collinge 1924, Ticehurst 1935). By the 1960s and 1970s this had fallen to about 17% (Glue 1977, Buckley and Goldsmith 1975). The contribution of field voles increased from around 20–25% to 40–50%. The importance of birds and wood mice also seems to have declined slightly. The earlier studies were done when harvesting methods were less efficient, when corn was stored in stackyards during the winter before threshing and most farms had root crops available over winter. Brown rats had a good food supply and were abundant around most farms (Leslie 1952).

3.5 Summary

Barn owls catch a wide variety of prey, including small mammals, birds, reptiles, amphibians and insects but worldwide they depend on a relatively small number of species. In almost all areas, terrestrial small mammals and in particular rodents are by far the most important prey. Across most of temperate Europe and North America, voles are the dominant species, especially the field vole, the common vole and the meadow vole. Shrews are usually significant secondary species and sometimes become the most important prey. Mice and rats of the family Muridae are the main food in most Mediterranean habitats and over much of the tropics and sub-tropics, in Australia and on many islands.

This includes species such as the house mouse, black rat, cotton rat, rice rat and multimammate rat. Gerbilles are important in the semi-arid parts of Africa.

Barn owls are more specialist, with narrower diets, in more productive habitats where small mammals are abundant and more generalist, with wide diets, in less productive, usually drier areas with low small mammal densities. In most areas the species composition of the diet varies seasonally and longer-term changes occur in relation to vole cycles or periodic changes in the abundances of particular prey. Diet also varies locally according to small scale variations in habitat.

4 *Foraging behaviour*

T hroughout most of its range, the barn owl seems to be mainly or entirely nocturnal. The absence of day-time hunting has been noted from direct observation and by the use of radio-telemetry in North America (Colvin 1984, Rosenburg 1986), Africa (Taylor, personal observation), Malaysia (Smal 1987, Duckett and Wells, personal communication) and Australia (Dickman, Predavec and Lynam 1991). Populations in northern Europe seem to be the only exceptions, although there may be other unreported examples. In Scotland and the north of

England, barn owls quite normally hunt during the day as well as at night both in summer and winter. Of course, in mid-summer, there is little real darkness so they have no choice but I have often seen them delivering prey to their young in the middle of the day especially when there has been prolonged heavy rainfall the previous night. When feeding chicks, they usually start hunting between 19.00 and 20.00 hours, giving about 3 to 4 hours of daylight foraging. The Scottish birds are very rarely mobbed by other species but in the tropics, and perhaps elsewhere, those that do venture out in daylight are persistently and aggressively attacked by a great variety of birds from starlings to crows, making hunting almost impossible.

The bird's nocturnal activity makes it very difficult to investigate the details of its foraging behaviour, necessitating the use of radio-telemetry. Even with the aid of this tool, it is often only possible to obtain very crude results. Bright moonlight can be very helpful. By observing in the light of evening, we have been able to obtain precise information on aspects such as foraging habitat selection and hunting methods for the Scottish birds but similar details from other areas are scarce.

Nocturnal foraging may be a problem for some researchers but for others it has presented a fascinating topic. How does the barn owl manage to locate and capture its prey at night? Thanks to the painstaking and elegant work of Roger Payne, Eric Knudsen, Masakazu Konishi and others in the USA we now know that they can achieve this entirely by the use of sound and much of the barn owl's anatomy and behaviour is modified to this end.

4.1 General foraging adaptations

Anatomy

Barn owls frequently hunt from perches, but they also spend a great deal of time on the wing, slowly quartering suitable habitat. Like other open-country, flight-hunting owls and raptors, they have relatively long, broad wings which give a high wing surface area to body mass ratio, thereby generating increased lift and enabling them to fly slowly and make abrupt turns without stalling. Low wing loading also makes it possible to carry heavy prey items at slow flight speeds, thus reducing the power output required and hence minimising energy consumption when transporting food to young in the nest.

The barn owl's flight is silent, both at frequencies audible to the human ear and at ultrasonic levels (Thorpe and Griffin 1962). This reduces the chance of being detected acoustically by small mammal prey and also improves the owl's ability to detect noises made by the prey while flight-hunting. Graham (1934) identified three structural adaptations that tend to reduce flight noise. Along the leading edges of the flight feathers, particularly the outer primaries, there is a very prominent, stiff, comb-like fringe which encourages laminar air flow and hence cuts noise production (Fig. 4.1). Similarly, along the trailing edge of the wing, the primaries and secondaries have a soft hair-like fringe which probably reduces turbulence where air flowing over the top and bottom of the wing meet. Lastly, the downy upper surfaces of the primaries,

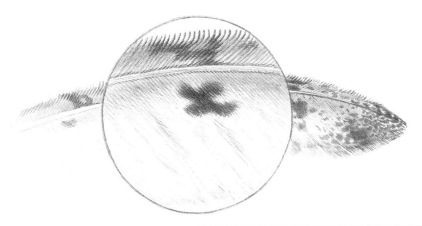

Fig. 4.1. Tenth primary feather showing serrated leading edge (enlarged).

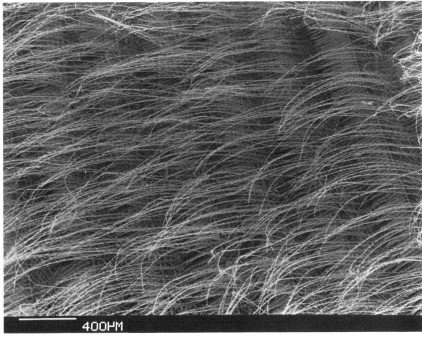

Fig. 4.2. The soft downy feel of the barn owl's plumage is produced by long, hair-like extensions of the barbules seen in this highly magnified scanning electron micrograph. These enable the birds to fly silently. (Photograph: Chris Jeffree, Edinburgh University Electron Microscopy Laboratory.)

400µM

secondaries and coverts must reduce noise that would otherwise be produced when these feathers move over each other during the normal wing beat. This downy appearance is produced by the extremely elongated extensions of the barbules which are illustrated beautifully in a photograph taken from a scanning electron microscope preparation (Fig. 4.2). It is significant that the fish-eating owls of the genera *Ketupa* and *Scotopelia* both generate considerable flight noise and lack the structural modifications described above.

The barn owl's legs are remarkably long and slender (Fig. 4.3), as are the toes and talons (Fig. 4.4), adaptations which make it easier to dive into long vegetation and give a wide spread at the point of capture.

Fig. 4.3. The barn owl's exceptionally long legs enable it to capture prey in long vegetation. (Photograph: Robert T. Smith.)

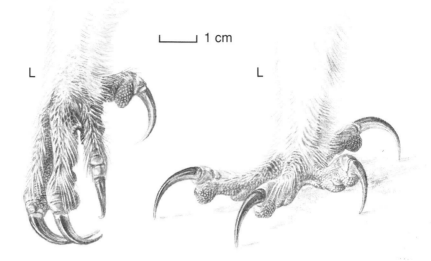

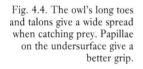

Fig. 4.4. The owl's long toes and talons give a wide spread when catching prey. Papillae on the undersurface give a better grip.

The undersides of the toes are developed into numerous papillae which probably help to maintain a secure grip on the prey.

Some of the barn owl's plumage patterns might also be important during foraging. The underparts vary greatly according to subspecies but it might be significant that the most northerly of these, *T. alba alba*, is the palest, as its name indicates. Most males are almost pure white over all of their underparts and whilst many females have some buff they are still pale in comparison with most other subspecies. Males do the bulk of the hunting for their partners and young during the summer which involves a great deal of daylight activity and winter day-time hunting may also be crucially important when prey availability is low and energy demands high. It is possible that their white coloration functions in the same way as the white underparts of sea birds or ospreys, reducing their conspicuousness to prey by minimising the degree of contrast against the sky. This might also be important on very bright moonlit nights. The upperparts of all races are various shades of buff with darker and lighter markings often giving a streaked appearance. When the birds are hunting from perches, which before the advent of fence posts would have been in low trees and bushes, they are probably highly vulnerable to predators such as larger owls or falcons approaching directly from behind. Their markings might again serve to camouflage them, this time against the background vegetation. The same argument might hold for birds roosting during the day-time in trees. The small Galapagos barn owl is extremely dark above, perhaps matching the dark volcanic rocks of the islands.

One of the most striking anatomical features of barn owls is their heart-shaped 'face'. This complex structure plays an important role in the owl's hearing and is discussed more fully below.

4.2 Vision

To my knowledge, the barn owl's visual abilities have not been investigated in convincing detail but thorough studies have been made of other owls, especially the tawny owl (Martin 1977, 1982). It would be inappropriate to extrapolate directly without qualification from tawny owls to barn owls as the former are adapted to hunt at very low luminance levels under the forest canopy whilst the latter, as open country hunters, would normally experience much higher light levels. Nevertheless, a brief description of the tawny owl's vision is useful. Martin discovered that the minimum amount of light detected by tawny owls (the absolute visual threshold) was about 2.2 times lower than that detected by humans. Although not great, the difference is significant as humans have rather good night vision. Compared with the pigeon for example, the tawny owl was about 100 times more sensitive. The lower threshold levels were achieved through the anatomy of the eye which is both extremely large and tubular in shape. In general terms, the lower the *f*-value of the eye, calculated by dividing the focal length by the diameter of the pupil, the brighter is the image produced which explains the advantage of large eyes. The maximum *f*-number calculated for the tawny owl was 1.3 whereas that for humans was 2.1. The greater the focal length of the eye, the larger is the image produced, and the better the spatial resolution. The tubular shape has probably arisen from a need to achieve a long focal length without a corresponding increase in overall eye weight. However, this shape results in a considerable reduction in the visual field and this means that to achieve good binocular vision the eyes must be more forward-facing than in birds with more rounded eyes. Despite all this, the spatial resolution achieved by owls is still poor at low light levels and certainly not nearly high enough for them to be able to catch prey by sight on the darkest nights.

The barn owl's eyes share some of the characteristics of those of the tawny owls, but are not as large, and it is likely that their absolute visual threshold lies somewhere between that of tawny owls and that of humans.

4.3 Hearing

The barn owl's ability to capture prey using only auditory cues was first established beyond doubt by Payne and Drury (1958) and reported in greater detail by Payne (1971). These researchers conducted trials in a light-tight room measuring 14 m × 4 m × 2.5 m, with two perches, 2 m high, at each end and the floor covered 5 cm deep in dead leaves. Owls were allowed to become accustomed to these surroundings in low light conditions. When light was completely excluded, the owls could still capture live deer mice and also a mouse-size wad of paper, pulled by string across the floor.

Details of the owl's behaviour were observed directly in the light and by the use of infra-red photography in the dark, and some clear differences were apparent. In the light, when a mouse was introduced into the room, the owl being tested would quickly turn to face it and then remain perfectly still for a while. Then, it would lower its head, lean forward with wings open, push off with its feet, take a single wing stroke and glide towards the prey (Fig. 4.5). During the glide its feet were tucked under the tail until about 1 m from the mouse, when they were brought forward just underneath the head. The head was then pulled back and the feet extended so that they took the lead, following exactly the same trajectory. At the last moment the talons were opened and spread wide. The mouse was captured with one foot whilst the other was used for landing.

In the dark, the sequence was the same until after leaving the perch, when, instead of gliding, the owl flapped very slowly towards the mouse, with its legs dangling underneath and not tucked back. In the final pounce, the talons were opened slightly earlier than in the light conditions.

Payne found that when fully spread, the talons of both feet lay equidistantly around the perimeter of an oval shape which, in the case of these north American barn owls, measured about 6 inches × 3 inches (15 cm × 7.5 cm). He also discovered that, in complete darkness, the owls were able to make adjustments just before impact so that the long axis of this oval and the body of the mouse were parallel, thereby maximising their chances of a successful capture. The owls were thus able to determine the direction of movement of the mouse.

To succeed in capturing prey this way, the owl must be able to determine the angle of the prey in azimuth (horizontal dimension) and elevation (vertical dimension) and also the distance to the prey. As yet, the mechanism for distance judgement has not been identified but it seems likely that learning and familiarity with the environment are involved. The ability to judge azimuth and elevation is based on the positioning of the external ears and the structure of the facial ruff (Knudsen and Konishi 1979, Knudsen 1981). When seen front-on the barn owl's head is actually pointed downwards so that the characteristic 'face' is formed by feathers attached to the top of the head (Fig. 4.6).

Fig. 4.5. Attack sequence.

Fig. 4.6. The barn owl's characteristic heart-shaped facial disc is part of their adaptation to capture prey by sound. This photograph shows a strongly marked female of the *alba* subspecies. (Photograph: Robert T. Smith.)

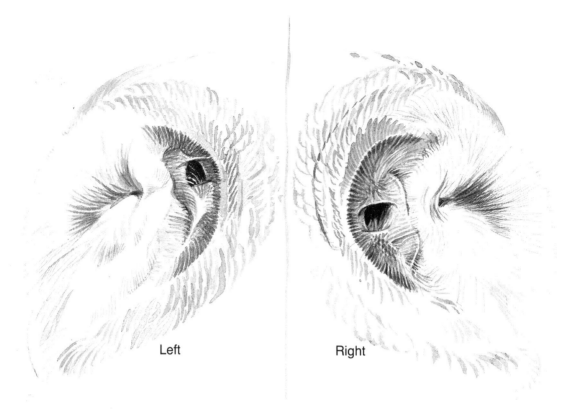

Left   Right

Fig. 4.7. Left and right ear positions, showing asymmetry.

The ruff is made up of layers of stiff dense feathers and is an effective reflector of high frequency sounds. Two concave troughs run along its length and these collect high frequency sounds and funnel them into the ear canals. The ear openings and the flaps of skin (the preaural flaps) that cover them are set asymmetrically (Fig. 4.7), the left being higher than the right. The orientation of the ruff also differs: on the left it is directed slightly downward and on the right slightly upward. As a result, sound reaching the two ears differs in many ways such as intensity, time of arrival and spectral quality, and it is by comparing these differences that the owl is able to judge both azimuth and elevation. The facial ruff amplifies sound and also, when it is removed, the owls are unable to judge elevation accurately (Knudsen and Konishi 1979).

The owl's precision is greatest when locating noises between 6 and 9 kHz and is poorer when presented with pure tones rather than noises containing a complex range of frequencies (Konishi 1973a). Natural sounds are rarely pure tones. Accuracy also depends upon how far the sound is presented to the side and below the owl. When presented immediately in front, the error is less than 2° in both azimuth and elevation which is a greater accuracy than shown by any other terrestrial animal so far tested including humans. However, accuracy decreases as the sound source is moved to the owl's side and especially when it is moved above or below. At an angle of 50° below, the owl's error is almost 20° (Knudsen, Blasden and Konishi 1979). This obviously has important implications for the optimum foraging height of the owl; with increasing height it will become less and less able to locate accurately single brief sounds made by prey. It also explains the importance of the initial head orientation response of the owl, bringing the target into its region of maximum angular resolution.

The barn owl also has remarkable skills in distinguishing between frequencies that differ by minute amounts. Both closely similar and complex noise spectra can be distinguished. Clearly, to do this the owl must be able to learn and memorise different sounds (Quine and Konishi 1974, Konishi and Kenuk 1975). This ability is essential if the owl is to distinguish between noises made by potential prey and those made by other components of the environment such as wind in the grass or non-prey species. It also opens up the possibility that they might be able to differentiate between various prey species and perhaps even between age or sex classes within species. As far as I am aware, this remains to be tested but a demonstration of the owl's full capabilities in this area would greatly assist our understanding of other aspects of foraging such as prey selection behaviour.

Barn owls are not only considerably better than humans in their ability to judge angles by hearing, they are also more sensitive to lower sound pressures at frequencies up to about 10 kHz. The owl is comparable with the cat up to 7 kHz but at higher frequencies, the cat can hear faint sounds inaudible to the owl (Konishi 1973b). The high frequency sounds the owl is best able to detect are exactly those made by prey when moving through vegetation, chewing or vocalising (Payne 1971).

The barn owl is thus extremely well equipped to locate and capture its small mammal prey in complete darkness, a skill which it must also put to good use when catching unseen prey through a covering of snow (Taylor, personal observation) or when taking voles concealed in their runways in long grassland. This does not mean that barn owls never use visual cues in hunting. Even in night-time hunting, their sensitive sight will be needed to move among potential obstacles such as branches. When hunting in daylight it is possible that vision is used alongside sound to locate prey. Voles are frequently seen by the human observer as they move between runway systems (Rice 1982, Taylor, personal observation) and species such as kestrel are able to hunt voles by sight alone.

Using the information presented in the above publications relating to the owl's accuracy, along with measurements of their talon spread and the size of prey, it is possible to estimate the optimum height at which they should scan and pounce for prey. For European barn owls, this works out at around 3 m, which corresponds with actual observations in the field (described later, p. 60). Payne (1971) calculated an absolute maximum limit of about 6 to 7 m for the much larger North American barn owls and found that his captive birds would not pounce directly from that distance but would fly closer and hover, presumably to obtain more accurate information.

4.4 Prey searching in wild birds

Barn owls are most conspicuous when they search in flight and this is often believed to be their main mode of hunting. However, hunting from perches is both widespread and frequent, many authors have noted its importance. In the oil palm plantations of Malaysia, this seems to be the main method, the birds habitually scanning from the side branches of the palms at a height of about 3 to 5 m (Lenton 1984a). This subspecies seems to have relatively short wings for its body mass (Chapter 2) and it seems likely it may have evolved in a more forested environment than other subspecies where shorter wings and a 'sit and wait' hunting technique would be more appropriate. Galapagos Island barn owls were observed to spend 72% of their hunting time on perches at heights from 0.5 to 2.0 m and 28% on the wing (de Groot 1983). Perched hunting was also the dominant foraging method in Texas farmland (Byrd 1982).

The owls regularly used fence posts along field boundaries for perch hunting in the Scottish farmland study area. However the searching method was found to be somewhat complex, making simple comparisons of the time spent perched and flying rather difficult. Perching phases lasted anywhere from about 15 seconds to about eight minutes and sometimes the owls pounced directly from the posts. More usually, however, they took off, rising to hover for a short period at heights of about 3 to 4 m before either making an attack or moving on to a new post. Often a series of high hovers was made, with only slight changes of position between them and sometimes a high hover was followed by a hover at about 2 m before the final pounce, as if the owl was attempting to obtain more detail first. Most often the owls simply gave up on one post and moved to another, but sometimes, while moving between posts, they pounced or hovered briefly. Therefore, although seemingly just transferring between perches, they were obviously still listening for prey. A clear separation of perched hunting and flight-hunting methods is obviously not easy. It seems likely that the perched phase was used for initial scanning, entailing the minimum energy expenditure, but that flight was often needed to pinpoint the exact location of the prey, both phases being integral parts of a single hunting technique which can be referred to as 'perched hunting'.

This method can be contrasted with the behaviour that is normally recognised as flight-hunting, in which the bird flies slowly using very shallow wing beats over open ground at heights usually of 1 to 3 m. In effect, this is the equivalent of the perched phase described above and, as in the perching technique, the bird can either pounce directly or hover before pouncing. The flight-hunting method clearly entails a greater energy expenditure to provide propulsion and also exposes the lightly feathered underwings. This large area has been shown by Dominic McCafferty of Edinburgh University to lose heat at a greater rate than most other parts of the body and, whilst this is probably not important during summer, in winter it could be significant.

Many other small mammal-eating birds of prey such as kestrels, also make use of perched and flight-hunting techniques. European kestrels experience a much higher prey capture rate from flight-hunting (Pettifor 1983, Masman, Daan and Beldhuis 1988) and there is a marked seasonal change with perched hunting being used more frequently in winter than in summer (Village 1983). The emphasis is thus shifted from a high cost but high return method in summer, when a high capture rate is needed to feed young, to an energy conserving, low return method in winter when they have only themselves to provide for.

Barn owls in Scottish farmland seem to follow something of the same strategy. From spot observations in daylight, the percentage of foraging owls using the perched technique increased significantly from 54% in summer to 87% in winter ($p < 0.05$). Part of this seasonal difference was associated with slight changes in foraging habitat selection: in summer the birds sometimes flight hunted over pastures and hayfields and sometimes even over cereals, particularly over patches where the crop had failed. Winter hunting was entirely along edges where perches were always available.

Habitat clearly has a major influence on the barn owl's choice of hunting methods. In conifer plantations, before the trees are tall enough to offer suitable perches, most hunting in summer is done in flight. Flight hunting must also have been the main method used by the New Jersey birds studied by Bruce Colvin and Paul Hegdal. Although they could not observe the birds directly because of their nocturnal habit, radio-tagged individuals in coastal areas hunted mostly over saltmarsh and crops such as lucerne where perches were not available.

It was shown earlier that the optimum height for prey searching was determined by the accuracy with which the birds could detect prey by sound. With increasing height, scanning area increased but accuracy of detection decreased. For European barn owls, the maximum range for accurate detection was about 3 to 4 m. This corresponds well with the hovering positions used by the birds, but they would often descend, probably after the initial detection of a small mammal, to about 2 m before pouncing, thereby, presumably, increasing their accuracy and success. I could not determine what heights were used during flight hunting in darkness, but, in daylight, the birds preferred heights between 1 and 3 m, which is lower than the height at which prey can be detected. However, when flight hunting the birds would often make an instantaneous pounce so they may have been sacrificing scanning range for quick and accurate prey location, perhaps taking advantage of surprise in conditions where prey may have been able to detect the owl.

When perched hunting, the birds generally had little choice as all the fence posts were about the same height (1 to 1.5 m). However, unlike kestrels that also hunted mainly for voles in the same area, barn owls were never seen to scan from telegraph posts. This emphasises the difference in prey detection range between visual and auditory hunters. Malaysian barn owls perch hunt from heights of 3 to 5 m (Lenton 1984a) but they have a greater talon spread and catch much larger prey than European owls and therefore can be successful at a greater range.

Earlier, I described how African barn owls will sometimes catch invertebrate prey such as termites by pursuing them on foot. We have also observed them apparently pursuing field voles in this way in Scotland and Lenton (1984b) described the use of this technique in Malaysian oil palm plantations. Although presumably overlooked to some extent, this method seems to be of minor importance.

We know almost nothing of the finer details of the barn owl's foraging behaviour, of how capture success and capture rates vary with season or weather conditions or in different habitats. The technical problems in this are enormous. In the darkness of a cold winter night, the observer is fortunate to see the birds at all let alone quantify these details. Doubtless, in time, the technology of radio-telemetry and remote sensing equipment will advance sufficiently for these aspects to be tackled.

4.5 Summary

Barn owls search for prey from perches and in flight. Their technique is based upon an acute sense of hearing that is so highly developed they are able to locate prey in complete darkness. The accuracy of the angular resolution of their hearing exceeds that of most other animals and they are also able to learn and distinguish between closely similar sounds. The 'heart-shaped face', characteristic of the species, is mainly concerned with collecting and channelling sound waves and detecting direction. Barn owls are silent in flight and possess many anatomical adaptations to achieve this. Silence is probably important to allow them to use their hearing to fullest advantage but would also not alert prey. They have long, broad wings giving a low wing loading which enables them to search slowly and also increases the efficiency of carrying heavy prey loads. Their legs and toes are exceptionally long which improves their ability to penetrate long vegetation and gives a wide spread of talons at capture.

5 The ecology and behaviour of the prey

Knowing what barn owls eat is a first step towards understanding how individual birds and whole populations are affected by variations in their food supply. The next step is to learn as much as possible about the behaviour and ecology of the prey species, particularly those aspects that might determine their availability to the owls: seasonal and yearly variations in their abundance, how their activity patterns vary with changing conditions, what their habitat preferences are and how densities compare in different habitats. The provision of prey-rich foraging areas around suitable nest sites must also be the foundation of all conservation projects.

Throughout the following chapters, various aspects of the biology of small mammals, the owls' prey, will be brought into the discussion and elaborated upon when appropriate and the purpose of this chapter is to introduce the main points that are of relevance to the owls. Several researchers have been able to study the owls and their prey simultaneously but there is additionally a great wealth of knowledge of small mammals that can give valuable insights into the problems the owls are likely to encounter while foraging. Some also point to useful areas of future research.

5.1 Temporal changes in prey numbers

Nearly all of the barn owl's main prey species have one very important characteristic in common: their numbers vary greatly with season and from year-to-year. On occasion their numbers reach exceptionally high levels that, more often than not, are soon followed by extremely low levels. Only exceptionally do barn owls experience anything like a constant food supply.

In the Scottish study area, I was fortunate in that Dr Nigel Charles, formerly of the Institute of Terrestrial Ecology, had started to monitor annual changes in field vole and common shrew densities in 1975. He retired just as my study began but the monitoring was continued by myself, Ian Langford and Aleem Chaudhary, aided by several undergraduate students up until 1991. This gives one of the longest runs of quantitative data on these species ever obtained in Britain. We used grids of treadle-operated traps set in vole runways on a number of sample areas each year. By doing a series of trapping-out exercises in fenced-off areas, Nigel Charles established that the index for field voles obtained in this way was very closely correlated with the absolute abundance of the animals. As the sampling procedure took about 2 weeks of intensive work each year, we were able only to monitor abundance in April, which was approximately the start of the owl's breeding season. However, in an independent study within the same area, Dr Pat Lowe, again formerly of the Institute of Terrestrial Ecology, monitored small mammal numbers in mid-summer. Comparing the two sets of data, the correlation between spring and summer vole abundances was high (R_s = 0.85, $p < 0.001$), so the spring index was also a reasonably good predictor of the food abundance for the owls later in the year when the young were being fed.

Annual changes in both field vole and common shrew abundance followed a distinctly cyclic pattern (Fig. 5.1). During the owl study period, this had a strict 3-year periodicity, with peaks in 1981, 1984, 1987 and 1990. Changes in vole and shrew abundance were strongly correlated (Fig. 5.2, $R_s = 0.83$, $p < 0.002$), so that when voles became scarce so also did shrews, a phenomenon previously recorded in Sweden (Hansson 1984) and Finland (Henttonen 1985, Kopimäki 1986).

Cyclic patterns in vole abundance with a periodicity of about 3 to 4 years seem to occur throughout much of the barn owl's range in

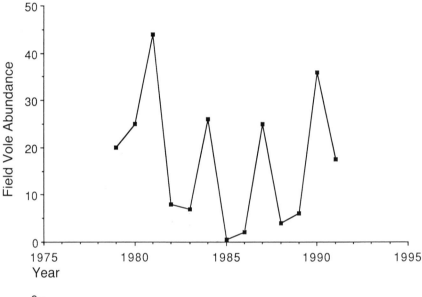

Fig. 5.1. The 3-year cycle in abundance of field voles in young conifer plantation habitat, Esk study area, south Scotland. Such cycles were synchronised in all habitats and at all altitudes within the study area.

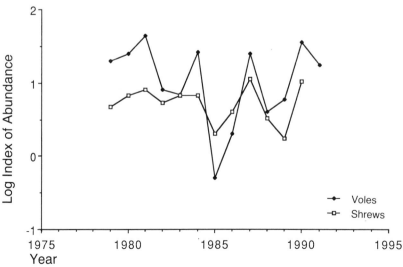

Fig. 5.2. Field vole and common shrew numbers cycled in synchrony in the Esk study area, south Scotland.

Europe, although there are some areas for which no information is available. Regular cycles, albeit from anecdotal information, have previously been recorded for field voles in Britain (Middleton 1930, Tapper 1979). For common voles, cycles have been recorded in Germany (Maerks 1954, Frank 1957, Bohnsack 1966), Holland (de Bruijn 1979, Wijnandts 1984) and Bulgaria (Straka and Gerasimov 1971).

Often these fluctuations are synchronised over large areas but many exceptions have been noted where nearby populations cycled out of phase (Middleton 1930, 1931, Chitty and Chitty 1962, Schnider 1972, Myllimaki, Christiansen and Hansson 1977, Hansson 1979, Tapper 1979). In my main Scottish study area (the Esk area, Area 1, Fig. 1.2, p. 4), in addition to the trapping, I recorded whether vole populations were at peak or low levels at numerous sample points each year, simply by examining the density of runways with fresh droppings. This showed that yearly changes in abundance were synchronised in all habitats and at all altitudes from just above sea level to 300 m. From 1981 to 1987, fluctuations in this area were synchronised with those in a second study area 70 km to the west (Area 2, Fig. 1.2, p. 4). However, populations declined in the first area in 1988 and remained high in the second area which experienced a decline the following year. Similarly, in 1991, the population in the Esk area declined but densities remained high in an area about 80 km to the north west. Such local variations could be important in the long-term population dynamics of barn owls and other vole predators.

The amplitude of the cycles seems to vary with habitat. Extensive grassland areas often experience very marked fluctuations (this study, Frank 1956, Hansson 1971b, de Bruijn 1979), whereas in more varied environments, changes seemed to be dampened (de Bruijn 1979, Village 1990).

In North America, the situation is more complicated not only as a result of the greater diversity of vole species (five to six species are important barn owl prey), but also by an apparently greater variation in the nature of their population fluctuations. Cyclic patterns have been described for all of the vole species eaten by barn owls, except perhaps for *M. oregoni*, but so also have non-cyclic changes (Taitt and Krebs 1985a). Indeed there is considerable debate over whether or not a real distinction exists between these (Sandell *et al.* 1991). For the owls, the difference is important; in a 3-year cycle, 1 year of good food supply is usually followed by 2 poorer years. Where voles are non-cyclic, low periods tend to be of shorter duration. The two systems should be expected to have very different effects on the breeding performance and population dynamics of the owls.

So far, nothing has been said of the causes of cyclic changes in vole numbers. What might seem a simple question has taxed the minds of many of the world's prominent ecologists for over 40 years and still there is no easy answer! A number of hypotheses have been advanced but the literature is so extensive and complex that it is impossible to give an adequate coverage here. The main ideas centre around changes in the quality and quantity of food available, predation levels and the animals' social behaviour. The reader who wishes to learn more should consult excellent reviews by Krebs and Myers (1974) and Taitt and Krebs (1985b).

The influence of weather on vole cycles was considered many years ago and largely discounted except for some species in particular environments. Indeed, Taitt and Krebs (1985b) did not even include this among the main current hypotheses. Shawyer (1987) asserted that field vole cycles in Britain are caused mainly by weather, particularly the duration of winter snow cover. He claimed that winters with 20 days or more of snow cover were followed by vole crashes and went on to conclude from this that long-term changes in owl numbers were caused by changes in weather patterns. The analysis on which these conclusions were based is open to serious criticism (Taylor 1992), but, nevertheless, the ideas have found their ways into some of the less rigorous British literature. Chitty and Chitty (1962) and later Krebs and Myers (1974) pointed out some of the main logical difficulties with simple weather hypotheses: populations experiencing the same weather can cycle out of synchrony, and *vice versa*, and cycles occur in a vast array of species living under completely different weather patterns, including some in areas where snow is rare or absent. In Scotland, field voles cycled in synchrony at all altitudes in the main study area, even though those at around 50 m above sea level experienced an average of only 13 days of thin, patchy snow cover whilst those at 300 m experienced an average of 35.6 days. Vole declines did not coincide with prolonged winter snow and high vole springs also followed winters when snow cover was considerably in excess of the 20 days suggested by Shawyer. There was no overall correlation between vole population changes between years and winter snow cover ($r = 0.15$, $n = 14$, *ns*, Taylor 1992).

Maerks (1954), in his analysis of many years' data for common voles could find no effect of winter weather other than an instance where an exceptionally mild winter seemingly extended a peak phase for 2 years. Prolonged extremely low winter temperatures not accompanied by the insulating effects of snow cover, as experienced sometimes in central and eastern continental Europe, might reduce common vole numbers (Frank 1957, Straka and Gerasimov 1971) but more research is needed to clarify this relationship.

A completely different aspect of weather, that of drought, seems to be important in the dynamics of some vole populations in areas where summers can be very hot and rainfall highly variable. Water shortage affects the growth of vegetation and hence the voles' food supply, which can result in reduced summer and autumn vole densities. This is important for meadow vole populations in New Jersey (Colvin, personal communication), for the California vole (Lidicker 1988) and for common voles in some parts of Bulgaria (Straka and Gerasimov 1971) and presumably some other parts of central Europe. In Scotland, field voles were most abundant in areas of moist grassland which remained succulent and green even when the highly drained surrounding farmland was showing signs of water shortage.

Seasonal patterns of vole abundance are highly variable, depending upon the species, whether or not the population is cyclic, and on the phase of the cycle in a particular year. Figure. 5.3 shows the types of change in the Scottish area established from regular assessments of the density of occupied runways. Populations were at their lowest densities in spring and increased from about mid-summer to reach highest densities in late autumn. During the summers preceding years with spring peak populations, numbers increased dramatically to reach the highest densities in the entire cycle by autumn. Declines occurred mostly in late winter and early spring, but, in peak years, there was less of a spring decline. It is particularly important to note the overwinter levels of vole abundance at different stages of the cycle for, as will be shown later, these have a marked influence on the owls' survival. At the vole peak, densities were high throughout the entire winter; in the following year

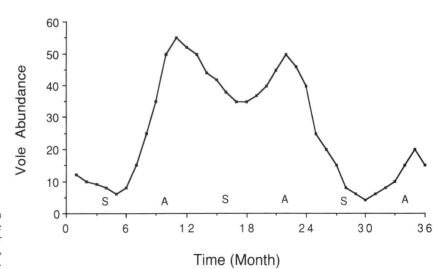

Fig. 5.3. Seasonal changes in field vole abundance over the course of a typical 3-year cycle, south Scotland. S, spring; A, autumn.

they were high at the start of winter but declined rapidly from January onwards and in the next year they were low throughout. Thus, during the 3-year cycle the owls experienced successive winters of increasing food shortage. Similar kinds of seasonal change have been described for common voles in Holland (Wijnandts 1984) and several North American vole species (Taitt and Krebs 1985a).

The other main barn owl prey species around the world share, or even exceed, the prolific reproductive potential of voles and show dramatic fluctuations in abundance but with no evidence of cyclic changes. Cotton rats and rice rats of the southern United States can increase and decrease rapidly (Otteni, Bolen and Cottam 1972) as can the multimammate rat of Africa. The latter is astonishingly fecund with litter sizes in Uganda, for example, varying from seven to 19 and averaging 12.6 (Neal 1968). It is often commensal, being found commonly around villages, and in agricultural habitats its numbers increase periodically to very high densities causing severe crop loss (Taylor 1968). Numbers also build up rapidly in areas of new growth following the burning of grasslands (Neal 1970). Generally, this species tends to breed in the latter part of the wet season and into the early dry season so that its numbers vary seasonally, reaching their annual high near the end of the rains and declining progressively thereafter.

House mice have long been known for their ability to increase under favourable conditions (Elton 1942, Evans 1949), and there is perhaps no more dramatic demonstration of this than the erratic 'plagues' of mice that cause so much agricultural damage, particularly in the southern states of Australia (Newsome and Crowcroft 1971, Newsome and Corbett 1975). More normally, on a smaller local scale, house mice also show considerable variation in numbers (Newsome 1969a,b). The causes of major outbreaks are not fully understood, but seem often to occur when there is a good crop yield and unusually high autumn rainfall, following a period of drought (Mutze, Veitch and Miller 1990).

Other important rodent prey of Australian barn owls, such as the plague rat *R. villosissimus* and the hopping mice *Notomys alexis* and *N. cervinus*, also fluctuate erratically in numbers and often at the same time as the house mice (Carstairs 1974, Newsome and Corbett 1975). Such changes are to be expected in rodents inhabiting arid areas and it seems likely that the normal situation for Australian barn owls, even before the introduction of house mice, is one of generally low prey abundance, punctuated by relatively short periods of exceedingly high densities, occurring unpredictably in time and space.

5.2 Habitat preferences

Most microtine voles are primarily adapted to live in a grassland habitat but they vary to some extent in the type of grassland preferred and in their ability to utilise crops. The majority of species, such as the field vole and meadow vole, are most numerous in areas of mature, moist grassland (Fig. 5.4). They construct complex runway systems within the grass or just below the surface and their nest chambers are usually close to the surface. They leave these runways to gather food which is mostly grass stems and leaves and a variety of dicotyledonous plants (Summerhayes 1941, Hansson 1971a,b, Evans 1973). Tussock-forming grasses and those producing large amounts of dead leaf material are important structural components of the vole habitat, especially during winter, and less fibrous, more nutritious species are preferred for food; swards with a varied composition of plants are most suitable.

In most farmland habitats adequate vegetation height for field voles is provided only in the edge areas of fencerows, hedges, waterways, road and railway verges and along woodland edges (Figs. 5.5, 5.6). Mature woodlands, cereal and root crops and silage are generally avoided although hayfields may provide good habitat for part of the year.

In Scotland, we did a series of trapping exercises to compare field vole densities in various farmland habitats. Densities in the rank grassland strips along woodland edges were much higher than in either the woodland or adjacent pastures or crops and many more voles were

Fig. 5.4. Areas of long, moist grassland are ideal habitats for voles and shrews. A fenced-off stream bank, south Scotland. (Photograph: Iain R. Taylor.)

Fig. 5.5. In farmland, small mammal habitat is often restricted to the edge areas of fencerows, ditches, hedges and woodland edges, south Scotland. (Photograph: Iain R. Taylor.)

Fig. 5.6. An excellent strip of grassland habitat for small mammals along the edge of a wood at the end of the growing season, south Scotland. (Photograph: Iain R. Taylor.)

trapped along woodland edges that had rough grazing areas adjacent to them rather than closely cropped pastures. Grass strips along hedgerows had similar densities to those along woodland edges (Table 5.1). Prey-rich foraging habitat for barn owls was therefore restricted mainly to these linear, edge areas.

Table 5.1. *The numbers of field voles trapped in various farmland habitats, Esk study area, south Scotland. May to July 1981. Each sample site had 20 trapping stations with two traps at each, at 10m spacing*

	Voles caught[a]
Low pastoral farmland habitat	
Coniferous woodland grass edge (2–4m wide)	14.2 ± 2.6
Deciduous woodland grass edge (2–4m wide)	12.7 ± 2.4
Hedge 'headland' (2–4m wide)	10.3 ± 2.2
Within coniferous woodland	0.5 ± 0.34
Within deciduous woodland	1.3 ± 0.7
Pasture	0
Silage	0
Hay	1.5 ± 0.6
Barley	0.2 ± 0.2
Sheepwalk habitat	
Coniferous woodland grass edge (4–6m wide)	30.5 ± 2.5
Rough grazing	2.7 ± 0.8
Within coniferous woodland	0.7 ± 0.3

Number of sample sites was six in each habitat.
[a]Voles caught are mean numbers caught over 5 days trapping per site ± S.E.M.

Hansson (1977) found that habitat utilisation by field voles was dynamic, changing according to the stage in the population cycle. At peak densities, the voles were distributed widely, occupying habitats of differing quality, but at the low phase of the cycle they were concentrated mainly in the best quality habitats. In farmland, it is probably important to have continuous networks of grassland to allow voles to disperse into and out of these optimum habitats.

Meadow voles in New Jersey farmland were also more abundant along edge habitats, although they also achieved quite high densities in some open crops, especially spinach and cabbage. However, highest densities were found in coastal saltmarsh areas, particularly in the salt hay communities (Colvin 1984). In some arid areas of North America and other parts of the world, irrigation ditches have rich productive grassland edges, encouraging high vole abundances in otherwise unsuitable areas (Marti 1988). At the other extreme, drainage ditches and canals in former wetland areas such as the fenlands of eastern England or polders of Holland often have wide grass strips that support high vole densities.

Common voles of continental Europe differ in a number of fundamental ways from the voles discussed above. This has very important implications for barn owls and their conservation. Instead of concealed surface runways, the voles dig underground burrows. In German studies, many of the females were found to live in communal groups in complex burrow systems which had nest chambers, food storage chambers, breathing holes and up to about ten entrances (Frank 1953, Boyce and Boyce 1988a,b,c). The burrowing habit frees common voles of the need for long grassland and they are thus found in a great variety of habitats including grazed pastures and meadows, orchards, hayfields, and crops such as alfalfa, clover, and some cereals but apparently not root crops to any great extent. They may reach their highest densities in regions where there are extensive, continuous grasslands, commonly referred to as 'plague areas'. In regions with more varied landscapes, densities tend not to reach such periodic high levels (de Bruijn 1979, Frank 1954, 1957, Goszcynski 1977). For continental barn owls inhabiting farmland, prey–rich hunting areas are therefore not confined to edge strips.

Most shrews, particularly those of the genera *Sorex*, *Blarina*, and *Neomys*, attain their highest densities in moist, moderate length grasslands and marshlands where they construct their own burrow and runway systems and also make use of those of voles (Spencer and Pettus 1966, Yalden, Morris and Harper 1973, Colvin 1984). The white-toothed shrews (*Crocidura* spp.) and shrews of the genus *Suncus* evolved in warmer climates and hence can be found in a wider range of drier environments (Lynch 1983). In the northern and western parts of its range in Europe, the common white-toothed shrew (*C. russula*) shows interesting seasonal changes in habitat. Often, it is closely associated with human dwellings and in a Swiss study it occurred mostly in dense vegetation in summer but in winter moved closer around buildings and compost heaps where the microclimate was warmer and invertebrate prey more abundant (Genoud and Hausser 1979). Perhaps this movement into more open areas explains why these shrews are taken by barn owls more frequently in winter, as discussed earlier (Chapter 3).

Mice of various species are found in a wide range of habitats. The wood mouse (*Apodemus sylvaticus*), for example, occurs in woodlands, hedgerows, long grassland, tussocky pastures and also in cereal and root crops such as sugarbeet (Flowerdew 1985).

5.3 Activity patterns

Most predators can locate active, moving prey more easily than static prey (Kaufman 1974) and activity patterns may therefore be important in determining the relative availability of prey to predators.

Small mammals, because of their size, are faced with a number of difficult problems. They have a very large surface area to volume ratio and this means that they lose heat in cold environments and gain heat in hot environments much more rapidly than do larger mammals. They also lose water more quickly through evaporation. Their rate of heat loss or gain is proportional to the difference in temperature between their bodies and the environment, and their metabolic rate and energy expenditure, per unit of biomass, are greater than those of larger mammals (Peters 1983). In relatively cold environments, they are therefore in a constant conflict between being active to secure a high rate of food intake and being inactive to conserve energy. In especially hot climates, the conflict is between foraging, conserving water and keeping cool. We should thus expect small mammals to show great sensitivity in their activity patterns to changes in the climatic conditions around them.

Shrews are right at the lower end of the size range and their energy needs are so high that they must forage more or less constantly to meet the demand. Voles such as the field vole, common vole and meadow vole, which depend mainly upon an energy-poor diet of grass stems, must also forage at frequent intervals. Shrews and voles feed throughout the day and night, alternating active foraging periods with inactive digestion and rest periods. Among voles, the periodicity of these active phases is related to their body weight, being just over 2 hours in field voles and common voles and almost 3 hours in the heavier meadow voles (Davis 1933, Brown 1956, Erkinaro 1969, Lehmann 1976, Daan and Slopsema 1978). Common voles and probably also the others show a marked tendency for activity bouts to be synchronised within local populations (Daan and Slopsema 1978). Thus for predators such as the barn owl there is a marked fluctuation throughout the day and night in the availability of voles. Indeed, in Holland, kestrels time their hunting to coincide with peak vole activity phases (Raptor Group RUG/RIJP 1982).

Superimposed upon these rhythms there are distinct diurnal patterns: activity is consistently higher in the early part of the night before midnight than later in the night and there is also consistently a resting period after dawn. Activity levels of voles during daylight hours are related to temperatures so that they tend to be more active during the day in winter than in summer. The amplitude of activity cycles is also much greater in winter, so periods of activity and hence increased prey availability are more pronounced. The voles perhaps shift their main foraging effort from night to day in winter to reduce the amount of heat loss. Especially on clear nights, radiation from their bodies to the cold sky would be very great (Davis 1933, Erkinaro 1969, Baumler 1975, Lehmann 1976, Daan and Slopsema 1978, Lehmann and Sommersberg

1980, Hoogenboom *et al.* 1984). As a result, day-time hunting is probably more profitable for the owls in winter than in summer.

Rainfall, particularly if associated with low temperatures, reduces vole activity (Baumler 1975, Lehmann and Sommersberg 1980). The chilling effect is exaggerated with wet fur: Frank (1954, 1957) noted that in winter, in conditions of extreme cold and also of high rainfall, common voles tended to remain within their burrows where they were able to use stored food supplies.

Many small mammals, including shrews, are more active on warm cloudy nights and less active on nights of full moons than new moons (Doucet and Bider 1969, Marten 1973, Vickery and Bider 1981). This could be interpreted as a behavioural adaptation either to reduce heat loss or to reduce predation risk.

Shrews also have diurnal foraging patterns, with higher levels at night than during the day and peaks just before dawn and after sunset (Churchfield 1984). Their activity shows many of the same relationships with temperature and cloud cover as other small mammals, but, interestingly, at least one species, the masked shrew of North America, and probably others as well increase their activity on rainy nights in summer (Doucet and Bider 1974). It may be that their invertebrate prey also become more active in moist warm conditions.

Winter is a very difficult time for shrews as their normally very high energy requirements would be further increased by low temperatures were they to behave as they do in summer. Many shrews, particularly those of the genus *Sorex* such as the common shrew, have evolved a very unusual method of coping with this: they allow their body weights to fall. Remarkably, this involves the skeleton as well as other tissues and the effect is to reduce energy needs very considerably. At the same time they also lower their activity levels by about a third and spend more time in their nests, again reducing energy requirements (Churchfield 1981, 1982). Shrews should thus be considerably less available to barn owls during the winter than in summer and also should provide less energy to the owls when caught because of their lower body mass. This seems at odds with the observations that shrews actually increase in the barn owl diet in winter. However, vegetation height is less in winter and this might offset the effects of reduced activity. Alternatively, there may, for some other reason, be a more profound change in the owls' prey selection behaviour in winter.

Many small mammals such as voles and shrews remain active under snow in winter as can readily be seen from the proliferation of surface runways when the thaw comes. A blanket of snow provides excellent insulation. Other species such as wood mice and rice rats are active on the snow surface; their footprints can easily be detected. Deep snow cover could thus alter the spectrum of prey available to the owls.

Another aspect of small mammal activity that should profoundly affect their availability to predators that hunt by sound, centres around their social behaviour. Many species, including most microtine voles and shrews, are territorial and especially in spring and summer are extremely aggressive towards others of their own species. Encounters between competitors are noisy affairs with much squeaking which is clearly audible to the human listener from several metres. Males usually occupy larger territories than females and are more mobile and active in defence which probably makes them considerably more vulnerable to predation (Clarke 1956, Turner and Iverson 1973, Myllymaki *et al.* 1977, Cody 1982, Churchfield 1990). The annual resurgence of sexual and territorial behaviour among the small mammals in spring should dramatically increase their availability to barn owls.

5.4 Summary

Most prey upon which barn owls depend vary seasonally in abundance and undergo very large year-to-year fluctuations. Among voles and shrews these changes are often cyclic with a 3- or 4-year periodicity but in some areas vole populations are non-cyclic. Other prey, particularly those inhabiting drier environments, show irregular and often dramatic fluctuations in relation to rainfall.

In temperate regions, mature, moist grassland is the preferred habitat for many of the owl's main prey, especially voles and shrews. Such habitat is usually only available in farmland areas as edge areas along fencerows, hedges, ditches, roadsides and woodland boundaries, although common voles of continental Europe, which dig burrows in the soil, occur more widely in pastures and crops.

The activity levels of small mammals vary greatly according to diurnal and seasonal patterns and also in response to weather. Such variations should profoundly affect their availability to barn owls.

6 *Prey selection, foraging habitats and energetics*

T he barn owl is often portrayed as an opportunistic, non-selective predator (e.g. Mikkola 1983). On a global scale, barn owls certainly exploit a wide range of prey species and weights and large differences can occur locally in relation to foraging habitat and prey abundance. They are specialists in that they depend mostly upon small mammals but obviously they also demonstrate a considerable degree of flexibility. It would be quite wrong, however, to take this as evidence of non-selective foraging. One would expect barn owls, like other predators, to have evolved ways of improving the efficiency of their predation above that obtained by a random approach to foraging.

Charnov and co-workers (Charnov 1976, Pyke, Pulliam and Charnov, 1977) formalised ideas on this subject in their optimal foraging theory, developed from the earlier work of MacArthur and Pianka (1966). In this, it is proposed that predators should forage in such a way that their net rate of food intake is maximised. Thus, this takes into account the relative profitability of different prey types that is determined by the time taken to locate them, the success of attempts to capture them, the time taken to handle or process them, the amount of food secured at capture and the energy expended by the birds in the process. The abundance of prey animals, their ability to detect and evade predators and their weights are all important elements in the equation. The theory predicts that when presented with a number of possible prey, predators should actively select the most profitable species, but that this selection should diminish as the absolute abundance of the most profitable prey decreases. In other words, as prey abundance declines and the time interval between encounters with individual potential prey animals increases, the predator should become more generalist and include a wider range of less profitable prey in its diet. Totally unprofitable prey should be avoided at all times. A wide range of bird predators have now been shown to follow these simple rules (e.g. Krebs *et al.* 1977).

During the breeding season the situation is further complicated by the fact that, in addition to meeting their own needs, the parent owls must make repeated feeding journeys to the nest. This behaviour is often referred to as 'central place foraging', and is clearly a time-demanding and energy consuming business. Many species improve their efficiency in this task by carrying to the nest only the heavier items they catch and by adjusting their load size to the distance from the nest (e.g. Kacelnik 1984, Kacelnik and Cuthill 1990). Energy and time constraints may dictate the maximum hunting range at this time.

In this chapter, I will look at the question of prey selection by barn owls, examining the evidence for selection among and within potential prey species in relation to their behaviour and body weights. The behaviour of parent owls when carrying food to the young will also be discussed.

6.1 Selection among potential prey species

The most detailed field investigation of selection among prey species to date has been made by Bruce Colvin (1984) in his New Jersey study area. He compared the owl diet at many nests with the relative abundance of each small mammal species estimated by trapping. His calculations of the relative abundance of these potential prey took account of their densities in each habitat type and the proportions of different habitats around the owl nests. Separate studies were made of inland farmland and coastal saltmarsh areas. In the former, white-footed mice formed 51% of the small mammal community in the wild but only 3% of items taken by the owls. By contrast, meadow voles were only 5% of the small mammal community but 70% of the prey taken. Colvin suggested that the apparent preference for voles might have arisen by the birds foraging more often in grassland habitat where voles were most abundant. However, even when comparisons were restricted to the grassland areas, voles were still captured much more frequently than would be predicted by chance alone.

A similar study in Colorado (Marti 1974) gave a mouse (*Peromyscus* spp.) to vole (mostly *M. ochrogaster*) abundance ratio in the field of about 8.7 to 1.0, but a ratio in the barn owl diet of only 0.57 to 1.0.

Comparison of the abundance of small mammal species is often susceptible to problems of differential trapability, but the differences described above are too great to be accounted for in this way. In areas where voles are available, barn owls seem consistently to show a strong positive selection for voles and a negative selection for various species of mouse (Fig. 6.1). Are voles therefore more profitable prey? In North America most species of vole are considerably heavier than most mice. Marti (1974) gave mean body masses of 21 g for *Peromyscus* species and 40 g for the vole *M. ochrogaster*. Colvin (1984) gave 40.7 g for meadow voles and 19.3 g for white-footed mice. The reward from the capture of a vole was obviously greater, but this alone cannot explain the magnitude of the selection for voles. Because of the much higher relative abundance of mice available to them in these study areas, the owls would still have achieved a higher rate of food intake by catching mice and voles roughly in the ratio encountered in the wild, rather than selecting so strongly for voles. Some additional factors must also be operating.

Mice are generally more active than voles. They have larger eyes and better hearing and are more agile in avoiding predators. The jumping mice (Zapodidae) of North America respond to disturbance by making a series of leaps of more than 1 m in length, then remaining completely motionless. Although common in many barn owl areas they are rarely taken in any quantity. Members of the genera *Peromyscus*, *Mus* and *Apodemus* are all capable of jumping to avoid predation. Barn owls might therefore prey selectively on voles simply because, once located, they are easier to catch than mice.

Fig. 6.1. Voles are probably the preferred and most profitable prey for barn owls in temperate areas. Large male voles are caught preferentially. (Photograph: Donald Smith.)

Attempts have been made to investigate this question experimentally using captive barn owls and these studies shed more light on the problem. Fast and Ambrose (1976) simulated a grassland-type habitat in a barn measuring 9×13 m and presented a single barn owl with equal proportions of white-footed mice and meadow voles. The owl caught many more voles than mice, in the ratio of 5.0 to 1.0 and, although limited in the number of birds and trials, the results supported the earlier field observations. Interestingly, the authors noted that the mice were noisier and much more visible than the voles as they moved around the trial area. It seems unlikely, then, that the owls were less able to detect the mice. Another, more thorough, set of trials was conducted in Virginia using three captive females (Derting and Cranford 1989). When presented with various combinations of meadow voles, deer mice and white-footed mice, the owls captured voles twice as frequently as mice, and voles were the first prey caught in 88% of the trials. Their success in attacking voles was 84% but only about 50% when attacking mice. If anything, the confines of the enclosure may have reduced the opportunity for the mice to avoid capture by jumping so capture success in the wild may have been even lower still. The time taken to subdue and swallow voles was slightly greater but this was considerably offset by the increased success of attacks and the higher body mass of voles, so that voles were much the more profitable prey. An almost identical set of results and conclusions were described from a very similar set of trials involving Ural owls in Japan (Nishimura and Abe 1988).

It is not clear from these studies whether the selection shown for voles in the wild is simply a result of a high capture success in comparison to that with mice or whether the owls choose to ignore the mice, making far fewer attempts to catch them. For this to occur, the owls would have to be able to distinguish between voles and mice before making capture attempts and as they hunt mainly by using auditory information this probably means that they would need to differentiate the sounds made by them. As far as I am aware, there is no evidence either way for this, although captive owls can distinguish between closely similar artificial sounds, so it seems likely that they could also distinguish between these species. There is clearly a need for more extensive trials with captive birds before the question of selection can be fully understood.

In the Scottish study area, I showed that barn owls living in farmland habitats increased the proportions of wood mice and common shrews in their diet when the voles reached the decline and low phases of their cycle. In the plantation and sheepwalk habitats where wood mice were not available, the owls caught increasing numbers of common shrews. The shrews averaged about 9.0 g, approximately one-third the weight of the average vole, and therefore offered a greatly reduced return per capture. As vole and shrew numbers declined in synchrony (Chapter 5) the owls did not switch to shrews simply because they remained abundant when the voles declined. It seems likely that because of body mass and behaviour differences, shrews and wood mice are less profitable prey than voles. The increased capture of these species as vole density declined is consistent with Charnov's (1976) prediction that the predator should include greater proportions of the less profitable species as prey abundance declines. If this is the case, we might expect that, at low vole abundances, the owls will not only experience a reduced overall prey capture rate but will also have a greatly increased energy expenditure during hunting brought about by tackling species that are more difficult and less profitable to catch.

The question of selection and diet diversity in relation to prey abundance and optimal foraging theory can be examined more widely. In the dry, hot desert habitats of the Great Basin area of North America, barn owls tend to concentrate around the more productive areas of irrigated farmland. Marti (1988) found a close correlation between the percentage of irrigated farmland around nests and the diversity of the owl diet. Owls living in the less productive more typically desert habitats where small mammal density is lower catch a wide diversity of prey. In the irrigated, productive areas where there is a greater availability of vole habitat and consequently high densities of voles, the owls are more specialist, concentrating on the voles.

On a global scale, in temperate climates where productivity is high and small mammals such as voles can reach high densities, barn owls often specialise on just a few species and sometimes even on a single species of vole. By contrast, in the arid areas of North America, Chile, the Mediterranean and Africa, the owls are more generalist taking a wide variety of small mammals, birds, lizards and insects (Chapter 3). These observations are, in a sense, natural experiments and suggest that barn owls may indeed behave as predicted by the optimal foraging theory, specialising more when prey abundance is high and becoming more generalist when prey density is low. Studies in Australia are particularly interesting in this respect. Two of them (Morton 1975, Morton and Martin 1979) involved areas where house mice had reached high densities following more productive periods brought about by rainfall. The owls specialised almost entirely on this single species. However in another study (Smith and Cole 1989) representing more typical arid conditions, as many as seven small mammal species made up 80% of the items caught.

The selection of prey species and diet diversity may therefore be dynamic phenomena, with the owls specialising when food conditions are good and diversifying when they are poor. Such changes are surely adaptive, enabling the owls to survive better in more marginal habitats and through times of food scarcity. The significance of this in the owl population dynamics and conservation will be explored in the following chapters.

6.2 Selection within prey species

In the Scottish area, I was able to compare predation rates on male and female field voles and on males of different body mass. In spring, when female barn owls are incubating or brooding young chicks, the male does all the hunting and the prey he delivers to the nest accumulates during the night-time foraging period (Fig. 3.2, p. 28). Most is consumed by the female and chicks during the next day but sometimes there is a surplus, which might be stored deliberately as an insurance against poor foraging conditions. Such surpluses have been found with barn owls elsewhere (e.g. Baudvin 1980) and in some other species such as Tengmalm's owl (Kopimäki 1981). By visiting nests immediately after the males' hunting sessions it is possible to identify, weigh and sex samples of the prey that he has caught. Data were collected in this way for six nests in 1982, a spring of declining vole populations and for four nests in 1984, a year of peak vole numbers. The birds involved all hunted over simple uniform habitats of 2- to 4-year-old sitka spruce plantations.

At the same time, vole populations in the areas around the nests were sampled to obtain data on sex ratios and weight distributions. The main problem in sampling these populations was to devise a method that gave a true measure of the relative abundance of the sexes and size classes and not just an index of different activity levels. This was done by using unbaited, treadle-operated traps that were sensitive enough to be triggered by the smallest voles and were set in the runway systems. Attempts were made to trap out all vole populations in sample areas, and trapping was continued until no voles had been caught for 2 days. Continuing for longer would have risked catching new animals, probably males, moving into the vacated areas.

Comparing the sexes and weight classes of voles caught by owls with those at risk in the wild population, it is apparent that the owls were selecting not only male voles in preference to females, but also large males in preference to smaller ones (Table 6.1, Fig. 6.2). It might be argued that these results do not demonstrate selective capture but merely that the male owls carried only the larger male voles to the nest and consumed most of the female and smaller voles themselves. However, this possibility can be ruled out as the lengths of the lower jaws of the voles in the male's pellets did not differ from those in the pellets of the females at the nest. The explanation of this selection behaviour is probably to be found in the different behaviour patterns of male and female voles. Both are aggressive and territorial in spring and summer, but males have much larger territories which are defended more vigorously (Chapter 5). They have to be more active and more mobile than females and disputes between males are frequent and noisy. Vocalisation by male voles increases dramatically in the spring (Turner and Iverson 1973) and, as the owls locate prey by detecting sound stimuli, it seems likely that the male voles were simply more vulnerable to capture than the females. Large males are more likely to be territory owners than smaller males (Turner and Iverson 1973) and consequently are more likely to be detected by the owls. Similar selective predation of male voles has been demonstrated in studies of Tengmalm's owls (Kopimäki 1981).

Numerous other studies have shown barn owls to prey differentially on various age or size classes of rodents and some have demonstrated no selection. When catching larger species such as brown rats, the owls generally take the smaller size classes. In New Jersey, the mean weight of brown rats trapped on farmland was 200 g but those taken by the owls averaged only 86 g. Rats up to 90 g formed only 27% of the wild population but 63% of those caught by owls. One individual was taken whose weight was estimated to be 290 g (Colvin 1984). Similarly, in

Table 6.1. *Comparison of predation rates by barn owls on male and female field voles in the spring and early summer, Esk study area, south Scotland. Males are significantly more vulnerable to predation than females*

	Capture rate (%)		Number	Significance of difference (*p*)
	Males	Females		
1981				
Caught in traps	42.9	57.1	49	
Caught by owls	89.7	10.3	58	< 0.001
1984				
Caught in traps	46.8	53.2	79	
Caught by owls	92.1	7.9	126	< 0.001

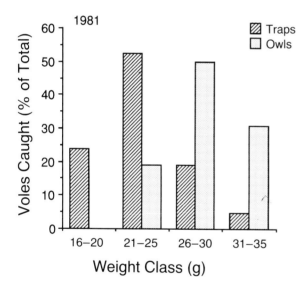 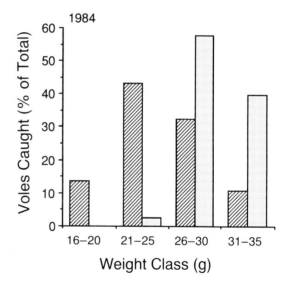

Fig. 6.2. The weights of male field voles caught by the owls compared with those caught by trapping in spring and early summer months. The owls caught more of the largest males.

Britain, barn owls catch younger brown rats averaging about 66 g (Bishop and Hartley 1976, Morris 1979) and in southern Africa they prey selectively on younger age classes of the multimammate rat (Davis 1959, Dean 1972).

Young rodents of these large species may be susceptible to predation through a lack of experience and attacking older animals may have involved the owls in an unnecessary risk of injury. However, there might also be an optimum prey weight range for other reasons, such as ease of swallowing or efficiency in carrying items back to the nest. The most frequently taken weight class of brown rats from both New Jersey and

Ohio was around 30–50 g which was also the range that included most average prey weights for the whole diet analysis within North America (Colvin 1984, Appendix 2).

An interesting study of predation on a population of house mice (*M. domesticus*) on Boulanger Island, Western Australia, showed that barn owls selected juvenile females. This component of the mouse population was forced by the adult mice into more open vegetation and independent observations revealed that the owls hunted more in these areas (Dickman *et al.* 1991).

Very young individuals of many species are often under-represented in the owl diet. In Scotland I hardly ever recorded young voles weighing less than about 15 g during the summer, even though these were the most numerous segment of the wild vole population. Similarly, in southern France most of the common voles and wood mice taken during July and August were adults whereas young animals were more abundant in the wild populations (Saint-Girons 1965). Predation on water voles is also concentrated on the older age classes (Kulczycki 1964). Common voles weighing from 5–15 g were largely ignored by barn owls in a Polish study area, whilst at the same time this size category made up the bulk of the voles caught by foxes (Goszcynski 1977). A probable explanation is that young individuals are less noisy than adults, not yet being involved in territory defence or sexual interactions, and that they therefore offer fewer cues to barn owls, whereas they would still be vulnerable to predators such as foxes that can detect prey by scent. The alternative explanation, that the owls avoided young voles simply because of their lower body mass and hence lower energetic reward, requires that the owls be able to distinguish age classes by the noises they make. European buzzards and kestrels that hunt by sight showed a significant selection for subadult common voles (Halle 1988).

Many predators are known to take substantial numbers of sick or diseased prey, which they find easier to catch. Surprisingly, given that whole prey items can easily be examined at nests, there has been almost no research of this interesting subject in barn owls. I have been able to find only one study which showed that individuals of the mouse (*M. spretus*) with bone deformities were more frequent in barn owl pellets than in trapped samples (Vargas, Palomo and Palmquist 1988). Presumably these animals were less able to move fast enough to avoid capture.

Many species increase the efficiency of their foraging when feeding mates or young by carrying the heavier items they catch back to the nest and consuming the smaller items themselves. To see if this was done by male barn owls during the breeding season, I examined five pairs where I knew the males to have only a single day-time roost site. I collected all the pellets produced by the males at these roosts and all those of the female and chicks during the 3-week period that the females were brooding their young in the nest. In this way, I could be sure that there was no possibility of confusing male and female pellets. The male provides all the food during this period. For four of the pairs, the prey consumed by the males did not differ from those they delivered to their females and young. The fifth male showed a slight but just statistically significant tendency to consume more of the shrews he caught himself and carry more of the field voles to the nest. The sizes of field voles (assessed by measuring mandible lengths in the pellets) eaten by the males did not differ from those delivered to the nest. There was therefore no consistent evidence that male barn owls select heavier items to carry to the nest.

However, they have another way of reducing their energetic demands when feeding their young. Males maintain a low body weight throughout the breeding season (Fig. 2.8) and one explanation may be that by doing this they minimise the energetic costs of flight. In addition to this, I found that they do not consume prey themselves until well into the night's hunting. Radio-tagged males took all the prey they caught during the first 3 hours or so of hunting back to the nest. Only after about six to eight items had been caught did they start to consume prey themselves. An average sized vole would have increased their body weight by about 8%, so that, had they eaten an item at the start of hunting, they would have had to carry a significant extra weight around for several hours thus increasing their energy costs.

6.3 Size of prey items

Intuitively, one might expect prey size to be an important factor in the prey selection behaviour of predators but, as has been shown, this must be considered alongside the behavioural characteristics and abundance of the individual prey species. There is reasonable evidence to believe that voles, for example, are preferred prey items for barn owls and that part of this preference is based on their behaviour as well as their large size compared with other potential prey such as shrews. On the other hand, observations of African barn owls feeding on emerging termites (Chapter 3) illustrate the point that exceptionally small items can be profitable at least for a short period, provided they are available in large numbers and little effort is needed to find and catch them.

Many factors must have been responsible for the evolution of body size in predators and the size of prey available to be exploited must surely have been important. Commonly, one finds correlations between the body size of closely related bird species and the size of prey they catch. For example, European tern species are graded in size and so also are the length classes of fish that they catch (Pearson 1968). A good correlation has been found between the body weights of terrestrial predators and those of their prey (Vezina 1985). Many bird and mammal predators take prey equivalent to about 10% of their own body weight, implying that there is some fundamental ecological significance to the relationship which at present is not fully understood. Schoener (1968) demonstrated a close correlation between body weight and prey weight among different species of raptors.

Initial impressions from lists of prey species might suggest that European and North American barn owls habitually take an exceptionally wide range of weights from small shrews of only 3–5 g to brown rats of over 100 g. A closer examination of the distribution of weights in these diets shows that the great majority of items fall within a much narrower weight range. Therefore in Colorado, 74% of items were between 20 and 50 g (Marti 1974) and in France 96% were between 10 and 30 g (calculated from Baudvin 1983).

Estimates of the average weights of prey taken by barn owls from different populations in Europe, North America, Africa, Australia, Malaysia and the Galapagos Islands are given in Fig. 6.3. European populations have average prey weights ranging from 12.8 g to 25.0 g with the most frequent being about 18–19 g. The lowest values are from populations in Belgium and Holland where shrews make up a high

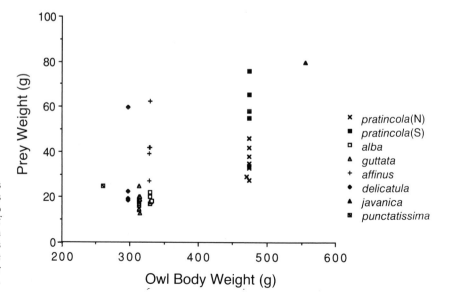

Fig. 6.3. Average prey weights taken by different subspecies of the barn owl in relation to the average body weights of male owls. Data from Appendices 1 and 2. There is a clear correlation, with the larger subspecies taking larger prey.

proportion of items and the highest values are mainly for diets with the greatest proportion of voles. North American averages range from 27.4 to 58.0 g, again depending mainly on the types of prey taken. Diets dominated by meadow voles show more uniformity varying only from 33 to 42 g. The highest values of 58 to 76 g are for populations in the southern states where cotton rats are important prey. African diets range from mean weights of 27 to 63 g, the lower values being from arid areas dominated by gerbilles and highest values from areas where the multimammate rat is the most numerous prey. Therefore, even within individual barn owl subspecies there was considerable variation in average prey weights, according to habitat and other factors.

When average prey weights are compared with owl body weights for the different populations, variation is again evident although there is clearly a good overall correlation (Figs. 6.3, 6.4). Using mean weights for male owls, which do not vary greatly through the year (Appendix 1), prey weights expressed as a percentage of owl weights for European populations ranged from 3.9% to 7.6% with an average value of 5.6%. For North American birds, the range was from 5.8% to 12.7% and an average of 9.0% and for African birds, 8.2% to 19.0% with an average of 13%. For the small Galapagos barn owl, de Groot (1983) gave a

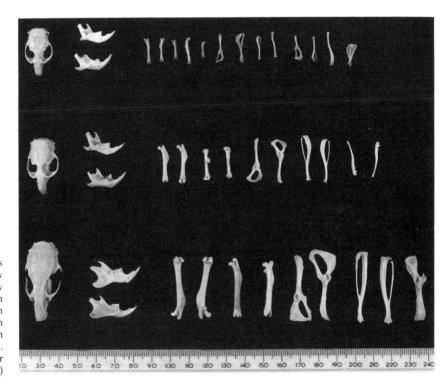

Fig. 6.4. Skulls and bones recovered from pellets show the range of standard prey sizes taken by different barn owl subspecies: field vole in Britain (top), meadow vole in USA (middle) and Malayan wood rat in Malaysia (bottom). (Photograph: Iain R. Taylor and Dave Haswell.)

mean prey weight of 25.0 g which represents 9.5% of the owl's body weight.

European barn owls seem to have exceptionally low average prey weights in comparison with the other subspecies. North American, African, Malaysian and Galapagos owls were reasonably close to the 10% of body weight suggested by Vezina, but the European birds were well below it. To be equivalent to the other subspecies, their average prey weights would have to be between 30 and 40 g. The owls cannot achieve this simply because small mammals of these weights are not available to them. Therefore, it could be argued that, in Europe, barn owl prey weights are below the limit that might be imposed by the birds handling and carrying abilities, and this is further supported by the large sizes of brown rats taken when the opportunity arises. Populations that are dependent upon shrews are very far below this limit. This could indicate that the foraging efficiency of European barn owls, especially when feeding young in the nest, is below that of other subspecies of the owl.

We have not considered so far how the weights of prey caught vary within populations of the owl through time and according to habitat. In the Scottish population, average weights showed a cyclic pattern over time corresponding to changes in the relative proportions of different prey species taken during the course of the vole cycles. The magnitude of the changes depended upon habitat: average prey weights of pairs nesting in conifer plantations changed from 14 to 22 g between vole lows and peaks, whereas for those nesting in the low pastoral farmland the range was only from 19 to 22 g. In the former the proportion of voles in the owl diet varied more over the vole cycle and the only alternative prey were common shrews but in the latter not only did the proportion of voles vary less but also the owls were able to switch to wood mice, which were heavier, in addition to shrews. Thus, in addition to experiencing a lower abundance of prey at low vole densities, the average weight of prey the owls were able to catch was also less. The efficiency of the owl's foraging was therefore considerably reduced during the low phase of the vole cycle. Not only must the frequency of their captures have been lower but also, once having made a capture, the weight and hence energy value of the items secured would have been less. The efficiency of transporting items to the nest would also have been reduced.

6.4 Selection of foraging areas

From what is known of the barn owl's anatomy, hunting methods and diet of small mammals, it is to be expected that its preferred foraging habitat would be open areas with moderate length vegetation. Furthermore, since voles, shrews and other grassland-living mammals form the bulk of the diet in all studies from temperate areas and other regions with relatively high rainfall or irrigation agriculture, we should expect that mature moist grasslands that provide the best habitat for these mammals would also be the preferred hunting areas for barn owls. Field observations from Europe and North America support these predictions and provide more precise details of habitat selection.

Generally, radio-telemetry is needed to guarantee unbiased information on habitat selection as direct visual observations tend to understate the importance of areas where perched hunting is used and over-emphasise observations of flight-hunting birds, which are easier to detect. However, unless radio-tagged birds can also be seen, even if only faintly, it is often difficult to pinpoint their position accurately enough to determine exactly where they are feeding. I attached radio transmitters to 23 owls in our Scottish area during the period from 1983 to 1985 and, together with Ian Langford, studied their habitat selection during the summer months in the farmland habitats. Our methods involved following individual owls continuously for up to 5 hours, and to achieve this two observers with radio receivers were always needed and sometimes as many as four were used. There was always enough light to see at least the outline of the birds and we were able to determine exactly where they caught their prey.

Potentially, the birds could have hunted over grazed pastures, silage and hayfields, cereal crops, and along field edges, fencelines, hedgerows, ditches and woodland edges. In practice, the pastures and silage fields were completely ignored and although the birds often made short flights over wheat and barley fields they caught hardly any prey there. A few items they did catch were from patches where the cereal crop had not taken and that had become colonised by arable weeds. All of the owls we followed caught most of their prey from edge habitats. Edges around woodlands were strongly preferred and hedgerows and fencelines were used much less than would be predicted from the lengths available (Table 6.2). Woodland edges consisted of irregular strips of rough grassland up to about 40 cm deep, varying in width from about 1 to 8 m (average 3–5 m) and separated from the adjacent crop by fencelines. The grassy strips along hedgerows, usually called 'headlands', had similar densities of field voles to woodland edges of similar width (Chapter 5), so the owls' avoidance of hedgerows cannot simply be explained by a shortage of prey. However, hedgerows offered few perching sites and could only be exploited by flight-hunting, whereas woodland edges were always bounded by fencelines and the birds did much of their hunting

Table 6.2. *Selection of foraging habitat by three radio-tagged barn owls in enclosed farmland, Esk study area, south Scotland. Woodland edges were strongly preferred. Six other radio-tagged birds for which large samples were obtained behaved similarly*

	Woodland edges	Hedge 'headlands'	Grass covered stream banks	Field crops
Bird 1				
Length of habitat within foraging range (km)	7.3	5.5	–	–
Locations of prey captures	34 (87.2%)	2	0	1
Spot locations of foraging birds	53	8	0	3
Bird 2				
Length of habitat within foraging range (km)	12.6	13.6	2.1	–
Locations of prey captures	51 (81%)	6	1	5
Spot locations of foraging birds	26	5	0	7
Bird 3				
Length of habitat within foraging range (km)	6.7	5.1	1.6	–
Locations of prey captures	29 (85.3%)	3	2	0
Spot locations of foraging birds	27	6	0	0

from the fence posts. This may therefore have determined their selection.

A strong preference for grassland and edge habitats has been established from radio-telemetry studies of the owls in several states of the USA. In New Jersey, 44% of the available habitat within 1.6 km of nest sites in coastal areas was grassland (mostly saltmarsh) but 79% of all radio locations of foraging owls were in this grassland. At inland sites, only 16% of the available habitat was grasslands (alfalfa, grass, hay and pasture) and 48% of foraging records were in these areas (Colvin 1984). Similar results were obtained in Virginia from the tracking of six birds (Rosenburg 1986). In Texas, pasture and light brushland, especially habitat edges along fencelines, were preferred (Byrd 1982). The researchers in all of these studies also examined the distribution of small mammal prey and found high concentrations along edge habitats.

6.5 Energetics

The ultimate purpose of foraging and prey selection is to secure nutrients and energy supplies, but at the minimum possible cost. Cost is usually thought of simply in terms of the energy and time spent obtaining food but the risks of injury and of predation by other birds of prey are probably also important. Individuals that forage more efficiently presumably are able to survive better during difficult conditions and to produce more offspring during the breeding season. Food requirements and the costs of foraging will vary depending upon the time of year, the bird's activity and prevailing climatic conditions. Eventually one would hope to quantify these gains and losses at least in terms of the energy involved, as this would lead to a more profound understanding of almost all aspects of the owl's ecology and behaviour, for example helping to explain in more detail why they breed at the times they do, why individuals differ in their breeding success and so on.

Unfortunately, this is a difficult subject to study with free ranging wild barn owls and progress is only just beginning. However, detailed investigations have been made of the ecological energetics of long-eared owls and kestrels in Holland (Wijnandts 1984, Masman *et al.* 1986, 1988) and these give a good idea of what we might expect to find in barn owls. These studies have demonstrated very large seasonal changes in the birds' food requirements and also differences between males and females. Wild male long-eared owls increased the amount of energy they used, or their metabolised energy (ME), from 211 kJ/day (equivalent to 40 g or two average sized common voles) in mid-winter, to 300 kJ/day (57 g or three voles) when feeding chicks. Male kestrels increased their demand even more, by 48%, in the breeding season compared with winter. The additional energy was needed to fuel the increased amount of flight and general activity involved in catching food for the young and carrying it to the nest. Females used much the same amount of energy as males during the winter months but much less during breeding, as they were less active.

Direct estimates of the daily food and energy needs of barn owls have been made only for captive birds, mostly of the North American subspecies. An average daily intake of 74.1 g of small mammal prey was recorded for two female owls (Wallick and Barrett 1976). This was equivalent to 444 kJ/day each and was provided by two average-sized meadow voles, amounting to 14.4% of the bird's body weight. Another female had an average daily intake of 60.5 g (Marti 1973). By subtracting the energetic value of the faeces and pellets produced by the birds and making appropriate allowances for small weight gains, Wallick and Barrett estimated that the birds actually used 360.4 kJ/day for metabolism. This is considerably higher than estimates of from 206.8–227.2 kJ/day obtained for four captive birds by Johnson (1974). However, the latter were kept under very tightly controlled conditions that allowed little movement, and fed non-live food, whereas the birds kept by Wallick and Barrett had to capture live prey in a large aviary. This illustrates a fundamental problem when captive birds are used to estimate energy consumption. Wild birds that have to search for their own food will obviously use more energy than captive birds but it is difficult to standardise the behaviour of captive birds during energy consumption trials and the corrections that need to be made in order to extrapolate to wild birds are inadequately understood.

The barn owl's zone of thermal neutrality, which is largely determined by the insulation properties of its feathers, is between 25 and 33 °C, which suggests that it is primarily adapted to relatively warm climates (Johnson 1974, Edwards 1987). In comparison, the thermal neutral zone of the snowy owl is from 3–18 °C (Gessaman 1972). Below 25 °C barn owls must alter their behaviour by fluffing up their plumage, crouching and so on and must also increase their metabolic rate in order to maintain their body temperature which is held consistently around 40 °C regardless of external temperatures. At 0 °C, the oxygen consumption and hence metabolic rate of inactive captive birds was 2.4 times that recorded at 25 °C (Johnson 1974). Six caged barn owls of the North American subspecies, averaging 561 g body weight, increased their energy consumption from 367 kJ/day to 560 kJ/day as the temperature fell from 25 °C to 5 °C. Their food intake (mice) increased from 77 g to 117 g per day (Hamilton 1985). This greatly increased energy requirement with falling temperature has profound implications for barn owls in the cold winter conditions of Europe and North America.

The problems faced by the owls in these conditions are currently under detailed investigation by Dominic McCafferty of Edinburgh University. Birds, in general, can gain heat from their own metabolism and also by absorbing radiant energy from the sun. They lose heat by convection through their feather layer to the surrounding air, by radiation and by evaporation. It is therefore particularly important to learn how the insulation properties, or conductivity, of the bird's plumage responds under different environmental conditions. By using model barn owls constructed from the skins of birds found dead, McCafferty has quantified the effects on heat loss of three important environmental factors: air temperature, windspeed and wetting of the plumage. The rate of heat loss experienced by birds with dry plumage at 0 °C and zero windspeed was estimated to be 2.5 times that at 20 °C. Insulation was reduced by almost 40% when windspeed was increased from 0 to 7 m/s and when the plumage was wetted this was reduced by a further 30%. Clearly, a barn owl hunting in cold, wet and windy conditions would lose heat very rapidly. Given a choice, one would expect them to forage in sheltered areas such as in the lee of woodland. One might also predict that when the weather deteriorates to a certain level, the birds should abandon foraging altogether. There is some evidence that they do indeed stop feeding during extremely wet conditions (Ritter and Gorner 1977). Long-eared owls also greatly reduce their flight activity during heavy rainfall and give up almost totally during periods of sleet (Wijnandts 1984).

Barn owls habitually roost in buildings and cavities during the day and from our radio-tracking in Scotland we discovered that they frequently rest up for shorter periods in farm buildings during their night-time foraging. This behaviour should enable them to conserve significant amounts of energy especially by reducing the effects of windspeed on convective heat losses and by lowering the radiation losses to cold surroundings. Dominic McCafferty has calculated that as much as a 30% saving in energy metabolism can be achieved in this way. This will be especially important when the birds are inactive, as at these times body heating cannot be achieved as a by-product of muscle activity, and also on clear winter nights when radiation to the cold sky would be greatest.

The minimum rate of energy expenditure of fasting inactive owls, termed the standard metabolic rate (SMR), was estimated to be about 144 kJ/day at 25 °C for three captive birds with body weights averaging 485 g (Johnson 1974). This is equivalent to a food intake of about 35 g of small mammal prey per day, or about 7% of the bird's body weight. To maintain themselves with no food intake, these birds would need to metabolise around 25 g of their own body tissue per day, equivalent to about 4–5% of their body weight. Johnson actually recorded a 16% body weight loss or 18 g/day in a bird that fasted for 4 days at 20 °C. At lower temperatures it would have needed to use more energy to offset the increased heat loss and thus would have used up more of its body tissues. European barn owls carry only limited fat deposits, amounting to about 5.5% of their body weight, compared with 10.0% for tawny owls and 13.4% for long-eared owls (Piechocki 1960). They must therefore start to metabolise muscle proteins soon after they begin fasting. Presumably the same applies to North American barn owls. Piechocki estimated that starved birds had lost 27% of their initial body weight (including fat). By comparing the weights of wild birds that had just died of starvation with those of live healthy birds, I estimated that male and female barn owls in Scotland lost 35% and 37% of their body weights respectively. Marti and Wagner (1985) estimated a 30% loss for North American birds. Using Johnson's estimates of SMR, the birds should not have been able to survive for more than about 8 days of total fasting at 20 °C and presumably less at low winter temperatures.

In December 1981, at some of the upland sites in my study area, a deep uniform blanket of snow formed within a few hours of a heavy snowfall that must have cut off the owl's access to field voles and common shrews which were the only potential prey species. Daily temperatures averaged around −3 °C and the snow persisted. Many birds perished and the first dead, emaciated barn owls were found 5 days after the snow fell. Surprisingly many of the owls nevertheless managed to survive these severe conditions. Presumably, some started off at higher body weights than others and some may have been able to catch some prey. However it is possible that many barn owls are helped through these difficult periods by utilising food caches they have stored earlier. This behaviour was first noted by Bunn *et al.* (1982) and I have observed it a few times, but John Cayford of the Royal Society for the Protection of Birds has recently confirmed that it is a regular aspect of barn owl behaviour in winter.

An alternative approach for estimating daily energy intake rates has been to quantify the prey remains in pellets. The assumption is made that only one pellet is produced per day and that the contents therefore represent one night's captures. This is supported by work on captive birds (Wallace 1948, Ticehurst 1935) but needs to be validated more thoroughly for wild ones. For two wild males studied over 30 days during November 1981 in Scotland, I obtained an average value of 75.2 g/day or 23.5% of the birds' body weight, equivalent almost to three average sized voles. A winter value of 73.8 g or 22% of body weight was obtained for one wild bird in Germany over 65 days (Knorre 1973). The Scottish birds survived the winter so probably were able to maintain themselves on this intake. Using the caloric value of European small mammals determined by Wijnandts (1984) and the assimilation efficiency of 77% given by Wallick and Barrett (1976) and Hamilton (1985) for captive barn owls, I calculated the metabolised energy of the Scottish birds to be, on average, 428 kJ/day. The average daily ambient temperature during this study was 4.5 °C. This value for metabolised energy is considerably higher than estimates of 280 kJ/day for the long-eared owl at the same time of year and at similar temperatures, even taking into account their lower body weights and may reflect the much thicker plumage and better insulation of the long-eared owl (Wijnandts 1984).

Estimates for Australian birds, which are about equivalent in body weight to European birds, are lower, 57.5 g/day (Morton and Martin 1979) and 65.5 g/day (Smith and Cole 1989) and this probably reflects the higher ambient temperatures in which these owls live.

6.6 Summary

In North America and Europe, barn owls frequently select voles in preference to other prey, probably because they are heavier and also easier to catch. The owls broaden their diet to include less profitable prey when the abundance of the preferred vole prey declines. Large male voles are captured more frequently than expected by chance, probably as a result of their greater activity associated with territory defence. The weight of prey items taken by different subspecies are related to their body weights.

In farmland, barn owls prefer to hunt along edge areas such as woodland edges and ditches where prey abundance is highest.

Barn owls have a thermal neutral zone of from 25 to 33 °C, suggesting that they evolved in warm climates. This has important implications for the species in northern areas in winter when they may frequently experience difficulty in obtaining their daily food requirements.

7 Ranging and roosting behaviour

Most owls are sedentary, spending all of their adult lives within specific areas. Even the so-called nomadic species such as the short-eared owl (*Asio flammeus*) must settle in one place for several months to breed and may spend long periods at one locality in other seasons. The way in which the birds use these areas, or home ranges, is of considerable interest to ecologists. What parts of the range are important for hunting? Do males and females share exactly the same range? Does range size vary with season or with food supply? Do ranges of adjacent pairs overlap? Defended home ranges, territories, can set limits to the density of some species and so can be important in the natural regulation of population size.

Because of the barn owl's nocturnal habit, details of ranging behaviour are difficult to obtain by direct observation and are best studied with the help of radio transmitters. These can be attached to the birds' tail feathers and can be recovered by catching the birds again or after the feathers have been moulted. Because of the technical difficulties involved, it is usually possible to follow only a small number of birds at one time and therefore there have been rather few extensive investigations of ranging behaviour. Much remains to be discovered.

7.1 Seasonal changes in ranging behaviour

Scottish barn owls do some of their hunting in daylight at most times of the year but especially so in mid-winter and also in mid-summer when feeding their young. In 1980, I marked 18 pairs with individually coded plumage dyes and, over the following 3 years, accumulated sightings of them obtained by chance while travelling around the study area for other purposes. Data collected in this way may contain biases but may validly be used to make simple seasonal comparisons within the same study area. From 1983 to 1985, a number of birds in part of the area were fitted with radio transmitters from late February to August which in addition to providing very detailed information on summer ranges also allowed a check on the data from the dye-marked birds. The two approaches gave very similar values of the distances from the nest at which birds foraged during the breeding season.

During the summer months of May, June and July, plumage-dyed birds were sighted up to 2.0 km from their nest sites with 89.5% of observations within 1 km of the nest. By contrast, in winter (December to February) the birds were seen up to 4.5 km from their nest with only 39.6% of sightings within 1 km, suggesting a greatly increased home range size (Fig. 7.1). In summer the birds centred their activities around the nest but in winter many of them, especially the males, associated

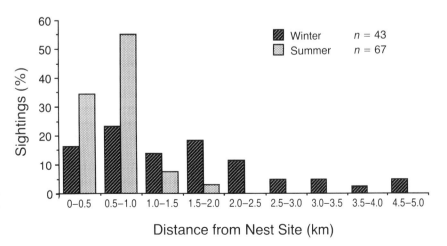

Fig. 7.1. Comparison of the sightings of dye-marked barn owls in winter and summer, Esk study area, south Scotland. The birds foraged significantly farther from their nests in winter (Mann–Whitney U–test, $z = -4.45$, $p < 0.001$).

more with roost sites that were often some distance from the nests. Therefore it might be argued that the apparent difference between summer and winter was not the result of changes in range size but rather of shifts in centres of activity. This was checked by reanalysing the data using only those birds (42% of the total) whose main roost was within 250 m of the nest and again the distances of sightings from the nest were significantly greater in winter than in summer. Therefore it seems that the birds genuinely ranged over much greater areas in winter.

The reason for this difference is not immediately apparent. In summer the maximum range may have been limited by the energetic constraints of transporting food back to the nest; some rich feeding areas may simply have been too far from the nest to be exploited economically. Perhaps in winter, being no longer constrained in this way, the birds simply wandered further to make use of such areas. Alternatively, it is possible that they were compelled to search over a much greater area in winter to locate patches offering suitable foraging conditions. Weather imposes a detailed mosaic of environmental conditions upon the countryside in winter and this could be extremely important to the owls. The depth of snowdrift depends upon wind direction and details of topography and vegetation. South facing aspects of woods, banks and hedges are much warmer than north facing aspects and snow and frost thaw, melt and disappear faster. The severity of rain, sleet and wind depend on the layout of suitable cover. All of these could affect the owl's ability to catch prey and determine the profitability of particular patches, forcing the bird to move around to locate the best areas. I have found no evidence of migration either regularly or in response to short periods of severe weather.

7.2 Home ranges during the breeding season

Using the radio-telemetry technique, Ian Langford and I studied six pairs of barn owls through the late winter and summer of 1983, a low vole year, and four pairs through 1984, a peak vole year. All pairs were in the low altitude pastoral farmland section of the study area. The method involved following individual birds continuously for periods of up to 5 hours and plotting their foraging paths directly onto maps in the field. The data collected were thus rather different from the usual spot locations obtained from most radio-tracking studies, although information of this kind was also obtained at other times. On some nights, by using four observers, males and females of individual pairs could be followed simultaneously, but generally the males were followed more than the females.

None of the birds used all parts of their range equally. At the time when most observations were made, they were feeding either their mates or their young in the nest, so foraging trips involved outward searches until prey were caught and return trips to deliver the prey. The birds clearly knew their ranges intimately and usually flew directly from the nest to favoured hunting areas which were mainly the grassy edges of woodlands, ditches and hedges (Chapter 6). Some tended to follow the same pathways each night from one foraging area to another. Return trips with prey normally followed the most direct route, usually at greater heights than the outward flights. Thus, the birds' flight paths frequently crossed crops and woodlands that were never used for foraging. This presents some difficulties when we try to find a suitable method for quantifying home range size. The simplest approach is to calculate the Maximum Polygon Area (MPA), that is, the area enclosed by a convex polygon joining up the outermost points visited by the birds. This would include within it all these areas that the birds did not hunt over but overflew and thus might seem to exaggerate the area actually used. However, the woodlands, fencerows and hayfields, although not used directly for hunting, may actually have been extremely important components of the range. The moist long grassland edge habitats, especially along woodland edges, that were the owls' favoured foraging areas were probably optimal habitat for voles and shrews and good habitat for wood mice (Chapter 5). Voles and shrews, especially, are territorial and it is likely that individuals removed by the owls from the edges were soon replaced by animals that were living in less good habitat in adjacent areas, or by non-territorial animals. Thus, woodlands, some fields and hedgerows and fencerows may have acted as reservoirs of prey, constantly topping up the edge areas used by the owls. Some crops may also have been important by enabling the dispersal of voles into otherwise isolated pockets of suitable habitat during the increase phase of the vole cycle. Therefore, unless one has a very detailed understanding of the dynamics of the entire ecosystem, it would be misleading to conclude that any part of the area encompassed by the maximum polygon range was unimportant. There is no sound scientific basis for subtracting seemingly unused areas from the calculation of range size.

Other methods for calculating range size generally assume that the home range is circular or eliptical and that the distribution of the bird's use of the area and hence the data points, follow particular statistical patterns. Most of the locations observed for the owls in this study were determined by the distribution of edge habitats, especially woodland edges, which conformed to no predictable statistical pattern. These more sophisticated methods of analysis might therefore simply have given a misleading picture. Therefore, although not ideal, I used the MPA method to estimate range size. Apart from anything else this also allows comparisons with other studies that used the same approach.

Estimates made in this way depend very much upon the length of time over which the data are collected. The Scottish barn owls were very active when feeding their young and moved quickly over their hunting range. One of the first birds we followed flight hunted over more than 50% of his total range within the first hour of observations. Generally, only 6 to 10 hours of continuous tracking was needed to determine the final range size.

Range sizes for the six males in 1983 were from 245 to 392 ha and averaged 319 ha, the equivalent of about a 1 km radius around the nest site (Table 7.1). Nests were mostly closer to the centre of the ranges than to the edges but many roost sites were right on the edges. Female home ranges overlapped almost exactly those of their mates, with only a few minor deviations, and did not differ in size.

The length of edge foraging habitat within the bird's range varied greatly among individuals; the preferred woodland edge habitat ranged from 3.4 to 12.6 km and total edges, including ditches, hedges and fencelines with grassy strips at least 1 m wide, varied from 5.2 to 26.2 km. No correlation was found between range size and the density of woodland edges or of total edges within the range. There was also no correlation between range size and the average distance from the nest at which prey were caught by the birds. Therefore, it seems that the owls did not increase their hunting range in poorer quality habitat nor did each pair take up a standard amount of hunting habitat. This seems surprising but can perhaps partly be explained by the distribution of habitats on the ground. For example, the pair on the range with the lowest density of edge habitat would have needed to fly about 1.5 km beyond the edge of their existing range in a southerly direction and over 3 km in an easterly direction to reach the next nearest suitable foraging areas. It might therefore have been more economical for them to search more intensively in a smaller range than to increase their time and energy costs by commuting up to 3 or 4 km from the nest to take in these new areas. The latter might only be feasible where an exceptionally profitable feeding patch can be exploited. This seems to be the case for some barn owls in the coastal areas of New Jersey where they may fly up to 5 to 7 km from inland nest sites on farms to feed over the prey-rich areas of coastal salt hay habitat (Colvin 1984, Hegdal and Blaskiewicz 1984).

There was no difference in maximum polygon range sizes between the peak and low vole year in the Scottish area but, on average, prey were caught closer to the nest site in the peak year (Fig. 7.2).

Table 7.1. *Home range size estimates for barn owls in Scotland and North America during the breeding season*

Locality	Range size (ha)				Source
	Individual males	Mean	Individual females	Mean	
Scotland	341, 248, 245, 352, 392, 333	318.5	327, 304, 275, 253, 376, 314	308.2	Taylor and Langford, unpublished
New Jersey	20, 809, 450, 196, 901, 3174, 662, 1414	953.3	621, 961, 224, 589, 1827, 72, 242, 376, 977, 632	652.1	Hegdal and Blaskiewicz 1984, Colvin 1984
Virginia	432, 193	308	151, 438	294	Rosenburg 1986
Texas	583, 252	418	223, 516	369	Byrd 1982

Range size is the calculated Maximum Polygon Area (MPA).

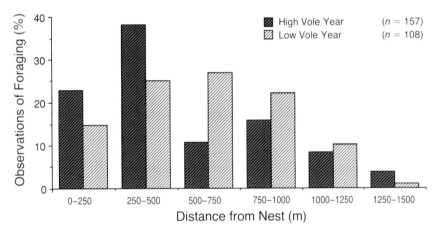

Fig. 7.2. Barn owls caught more of their prey closer to the nest in years of high vole abundance. Data from 17 radio-tagged birds, enclosed low farmland, Esk study area, south Scotland from 1983 to 1985.

Five of the pairs tracked in 1984 were nearest neighbours to each other and their ranges overlapped considerably, with no indication of defence of hunting areas (Fig. 7.3). Neighbouring males were sometimes seen foraging, with no apparent aggression, within about 150–200 m of each other in these overlap areas in circumstances which suggested they were aware of each other's presence. Birds returning to the nest with prey sometimes flew directly over their foraging neighbours without eliciting attacks or threats. Attacks were seen on only three occasions when intruders actually entered occupied nest sites. One of these cases involved a pair which were not neighbours and which tried repeatedly

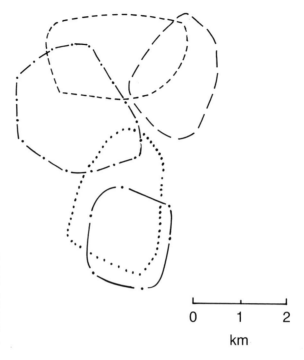

Fig. 7.3. Summer home ranges of five adjacent breeding males in enclosed low farmland, Esk study area, south Scotland, 1984. Average range size was 320 ha and considerable overlap was evident among neighbours.

over 2 hours to enter a nest building, despite the vociferous and violent attentions of the resident male and female. Overlapping ranges of adjacent pairs have also been recorded in New Jersey (Colvin 1984).

One of the farmers in my study area, who was a reliable natural historian, reported two pairs of barn owls nesting about 25 m apart in his farm buildings in a peak vole year before the start of my study. He often observed all four adults feeding together on a particularly rich damp meadow of only about 0.5 acres (0.2 ha) with no obvious aggression among them. Similar close nesting and apparent sharing of hunting areas has been reported from Utah (Smith *et al.* 1972).

The evidence we have available suggests that barn owl foraging ranges are either poorly defended or not defended at all, whereas the immediate nest area is definitely defended. Whether or not a foraging area is defended and to what extent must depend on many factors but probably is determined mainly by the relative costs and benefits of defence compared with non-defence. The costs of defence include the time and energy expended that could otherwise be used for prey searching, mate guarding or other activities that might improve an individual bird's fitness. Defence would bring little or no advantage if the bird's prey capture rates were not reduced in the short or long-term by the overlapping foraging activities of other birds or if the costs of defence exceeded the gains from improved food intake resulting from defence. The short-eared owl in temperate areas feeds in much the same way and on almost exactly the same prey as the barn owl yet defends hunting areas vigorously (Lockie 1955, Clarke 1975). Short-eared owls are diurnal foragers so the detection of intruders is presumably easier for them and they also take up remarkably small territories (e.g. 17.8 ha, Lockie 1955; 15 ha, Holzinger, Mickley and Schilhansl, 1973) so that the costs of defence must be relatively low. By contrast, it would be difficult for nocturnal barn owls to detect intruders over much larger areas. Also in many areas, suitable nesting places such as cliffs or riverside trees may traditionally have been more patchily distributed and more limiting than food supply. In these circumstances one would not expect the birds to have evolved a rigorous defence of feeding areas.

Range size among different species of raptors is related to their body mass: larger species occupy larger areas (Schoener 1968). Information is available for range sizes during the breeding season from three areas in North America allowing comparison of the much larger North American barn owls with the Scottish owls. Males of the latter subspecies have a body mass only 72% of that of the American subspecies (Appendix 1). Range sizes for the American birds were established from radio-telemetry by making point locations of foraging birds rather than continuous following as in the Scottish study. Only those ranges calculated from at least 30 locations were included in the comparison. For the Scottish birds, asymptotic range sizes were used from the plotting of cumulative range size against hours of observation.

Despite these attempts to standardise the data, a great deal of variation was evident, especially among the North American estimates (Table 7.1). In New Jersey, the size of the largest foraging range was some 20 times that of the smallest. It is possible that such differences were simply the result of inadequate sampling, but it is also possible that they were genuine as the nature of the habitat and the quality of the foraging areas varied greatly. Some birds covered very large areas, travelling out into coastal marshland (see above). Average New Jersey foraging ranges were much larger than those in Scotland. Ranges in Virginia were smaller and those in Texas larger, but not nearly by so much as those in New Jersey. It would be unwise to attempt any simple conclusions given the extent of these variations, and they serve to emphasise the need for much more research into the factors that influence range size.

7.3 Roosting behaviour

Barn owls use a great variety of roost sites both during the day-time and for shorter breaks during night-time hunting. In Britain these include stone chimneys, disused buildings, haysheds, cavities in cliff

Fig. 7.4. Barn owls often roost in very small cavities such as the spaces between bales in haysheds. Thus, they are able to conserve energy by reducing heat loss. (Photograph: Iain R. Taylor.)

Fig. 7.5. A small hay barn in a Pyrennean valley used by roosting barn owls. (Photograph: Iain R. Taylor.)

faces and tree holes which are often much smaller than those used for nesting (Figs. 7.4, 7.5). Pellets are often found in farm buildings such as cattle sheds where an owl has spent an hour or two resting during the night but where owls are never seen during the day. Where buildings are not available, they commonly roost in small woods or densely-spaced conifers. American barn owls roost in a similar range of sites but also use abandoned agricultural silos and have even been found roosting on the ground in crop fields (Colvin 1984, Rosenburg 1986). Over much of continental Europe, church towers are important roost sites (Baudvin 1986, Muller 1989).

During 1980, I attempted to locate all of the roost sites used by 18 pairs to whose plumage I had applied small areas of colour dye. The birds could be identified at their roosts without disturbing them and also their moulted, dyed feathers could be found at places where they were not seen. In 1983 the same was done for six radio-tagged pairs in the same areas. Apart from a few weeks spent together during egg laying, males and females roosted apart during the breeding season in 16 out of the 18 dye-marked pairs and in all of the six radio-tagged pairs. Each had from one to three alternative day-time roosts but all used one roost consistently more than the others. These main roosts ranged from a few metres to up to 2.25 km from the nest. Occasionally, birds were found roosting up to 3 km from their nests in summer. Eight of the dye-marked females and only two of the males had their main roosts on the same farm as the nest site but in a different building. Overall, males roosted significantly further from the nest than did the females (Fig. 7.6).

Summer roosting behaviour has been examined in several parts of the USA, using radio-tagged birds. Individual owls in New Jersey had from one to seven day-time roosts at distances up to 3 km from the nest. 65% of 120 roosts located were in woods and 25% in farmsteads or other buildings (Colvin 1984). In the same areas, Hegdal and Blaskie-wicz (1984) recorded birds roosting regularly up to 5 km from the nest with a maximum of 8 km.

Every time I visited barn owl nests throughout my study areas in the day-time, I recorded which of the adults were roosting there and this source of information also revealed a marked seasonal change in the bird's roosting behaviour (see Fig. 11.2). During the winter months about 40% of females and 15% of males roosted at their nest sites. However, as the breeding season approached, first females and then males spent more and more time at their nests. About 40–60 days before laying, there was a sudden increase with about 80% of females roosting at their nests and by about 10 days before laying all of them were roosting there. Males moved in somewhat later and by about 10 days before laying almost all were roosting alongside their mates at their nests. Females did all the incubation and brooded the chicks until they were about 3 weeks old, so remained on the nest for all of this period. However, more or less as soon as the clutch was completed, most males moved to roost away from the nest. Some remained within the same building but no longer roosted alongside their females. By the time the

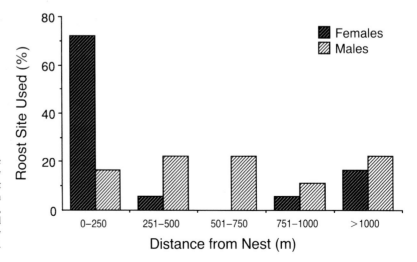

Fig. 7.6. Distances from the nest of the most frequently used summer day-time roost sites of 18 dye-marked barn owl pairs, Esk study area, south Scotland. Males tended to roost much farther away from the nest than females.

young were about 50–70 days old, all males and all but a few females were roosting away from the nest. At this stage the young could fly and were still either at the nest site or close by and the adults may have roosted away to escape their food begging. Many returned to roost at the nest after the young began to disperse at about 11 to 14 weeks of age.

Roost sites seem not to be defended consistently during the winter months; over the years I recorded many instances of up to four females sharing a roost and of two males sharing. By catching the birds I was able to establish that, in most cases with both sexes, the cohabitants were the resident males or females with first year 'visitors' who were not their offspring. Sometimes the visitors stayed for several months but more often they moved on again after several weeks whilst the resident birds remained. During food supplementation trials in an old stone barn through the winter of 1981, I recorded five individuals coming to the site to collect food, the resident pair who bred at the site, a neighbouring pair and a single unidentified bird. By contrast, I have never recorded two males sharing a roost site during the summer months. In those cases where there had been a long-term sharing of roosts through the winter, the first year bird always vanished sometime during March or early April. Aggressive interactions over roost sites, if they occurred, would probably have been short and infrequent and I never observed them, but it seems likely that males were indeed intolerant of each other at roosts during the breeding season.

7.4 Summary

In farmland habitats in Scotland, barn owls have a breeding season home range of about $3.2 \, km^2$; most prey are caught within 1 km of the nest. During winter they range more widely, up to 5 km from their nests. Breeding season hunting ranges of adjacent pairs can overlap extensively with no rigorous defence, but nest sites are strongly defended. Instances of close nesting and shared foraging areas have been recorded in several other studies. No difference in overall range size was recorded between high and low vole years but the birds caught more of their food closer to the nest in high years.

Barn owls roost in a variety of sites, including the attics of disused buildings, stone chimneys, haysheds, cavities in cliffs and trees, and under the canopies of coniferous trees. During the breeding season they can have several, alternative day-time roosts but if undisturbed tend to use one site consistently. Males tend to roost further away from the nest than females.

8 *Moult*

Bird feathers are, without doubt, one of the great products of evolution. They are lightweight but robust and so constructed that for the most part they can be kept in a functioning state simply by regular preening. Nevertheless, wear and tear is inevitable: it has been estimated that long-eared owls lose about 19% of their total plumage weight just between November and May (Wijnandts 1984) and that bullfinches lose 30% in the course of the year (Newton 1967). The reduction is a consequence of abrasion, breakages and loss of feathers. Inevitably, the bird's ability to fly efficiently and the effectiveness of its body insulation would be threatened and so regular replacement of feathers is necessary.

Moult is a major event in a bird's annual cycle and, like breeding, is usually undertaken at a particular time. It requires a significant increase in energy expenditure: most studies have estimated a value of 5–30% above resting level (West 1968, Chilgren, in King 1980). Much of this extra demand arises not from the production of feather proteins but from increased heat loss as a consequence of a greatly enhanced blood supply at the skin surface and the simultaneous thinning of the bird's insulation layer.

This all raises a number of very interesting ecological questions. What determines the timing of moult and how is it related to the seasonal abundance of prey? Does moult conflict with breeding? Is the pattern of feather replacement modified according to the bird's method of flight or hunting? In this section, I will explore what is known and unknown of barn owl moult and attempt to answer some of these questions.

8.1 Sequence of moult

The moult pattern of European barn owls was first described in Germany by Piechocki (1974) and Schonfeld and Piechocki (1974) from wild birds and captive birds kept in outdoor aviaries. By using small spots of dye on the wing and tail feathers of a sample of wild birds studied over 8 years, I was able to confirm that barn owls in Scotland follow much the same pattern as the German birds and, independently, the same pattern was discovered in North American birds (Bruce Colvin, personal communication). The only study of moult reported for barn owls in the tropics demonstrated a broadly similar sequence of feather replacement but over a very different timescale (Lenton 1984b). The sequence of barn owl moult is complex and differs from that of most other owls reinforcing the separate systematic position of the Tytonidae.

In temperate areas, replacement of the primary feathers begins in the bird's second calendar year of life and is not completed until the fourth calendar year. Moult starts in the centre, at P6 (Figs. 8.1, 8.2) and proceeds ascendantly and descendently. Usually only P6 is shed in the first moult period, in the bird's second year, but sometimes P7 is also replaced. At the second moult, P5, 7, 8 and 9 and sometimes also 4

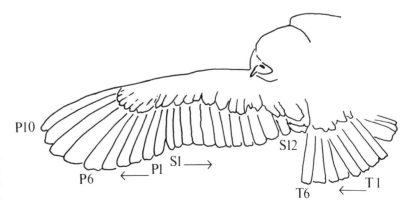

Fig. 8.1. The system for numbering barn owl wing and tail feathers. Primary moult usually starts with P6.

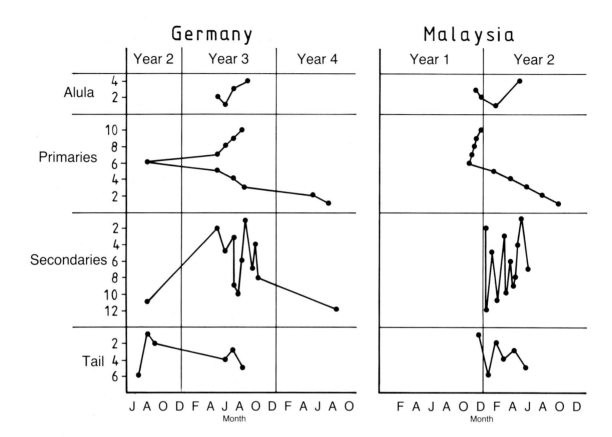

Fig. 8.2. Comparisons of the moult patterns of barn owls in a temperate area (Germany, Piechocki 1974) and in a tropical area (Malaysia, Lenton 1984b). The tropical birds were able to replace all of their juvenile feathers in a single moult period starting at about 300 days of age, whereas the temperate area birds took three moult periods, starting at about 400 days and spread over 3 years, to replace their juvenile plumage.

and 10 are renewed with the remainder normally being replaced during the third moult period in the bird's fourth year. About a third of the birds start a new wave of moult in the fourth year by again replacing P6. From the fifth year onwards, primaries are usually replaced in a 2-year cycle, centred on P6.

The moult pattern of the secondary feathers is much more variable. From one to four feathers may be shed in the bird's second year, usually S12 and a variable number from S11 to S14, with the majority of the remainder, (anywhere from six to ten feathers) being replaced in the third year. Any that are left over are shed in the fourth year. The four feathers of the bird's alula are replaced in their third and fourth years.

Tail moult begins in the second year, starting with the central feathers (T1) and the outer pair (T6). Following this there seems to be considerable variation with little indication of a regular pattern. Often one or two more feathers are shed at the first moult and then whatever remains is replaced the following year, the bird's third year.

The moult of body contour feathers and coverts has not been

described to my knowledge. However, in 1979 and 1980, I marked a large number of birds with plumage dyes on the under body and this colouring was still clearly visible on all survivors 2 years after application and on some even a year beyond that. It seems likely, then, that these feathers, like those of the wing, are replaced on a 2- or 3-year cycle.

Birds in the tropical climate of Malaysia moult in much the same sequence as temperate region birds, with minor differences, but, remarkably, replace all their wing and tail feathers in a single moult spread over a period of about 7 months. Moult starts when the birds are still in their first year, at an average age of 301 days (Fig. 8.2).

Temperate and tropical birds shed and replace most of their distal or outer primaries rapidly, more or less as a unit, whereas inner primaries and secondaries are renewed more slowly over a much longer time period. Also, secondaries tend to be shed in a sequence that alternates from one side of the series to the other, thus reducing the size of gaps and balancing them across the inner wing. Lenton (1984b) found that outer primaries, despite their greater mass per unit length, grew faster than inner primaries, secondaries or tail feathers. As the outer wing is largely responsible for propulsion and the inner wing for generating lift, this suggests that barn owls can tolerate reduced acceleration and flight speeds but attempt to restrict the duration of impairment, whereas a reduction in lift is minimised at all times. This is consistent with the owl's method of hunting, involving, as it does, slow flight speeds and carefully controlled, often gliding pounces, and the need to carry relatively heavy prey loads. A significant reduction in lift would mean a substantially increased energy expenditure when carrying food to young in the nest.

8.2 Timing of moult

Many diurnal raptors and some owls moult all their flight feathers each year and tend to follow a set pattern of replacement. With a knowledge of the growth rates of feathers, it is thus possible to estimate the starting date of moult for individual birds trapped at almost any stage during the moult (Newton and Marquiss 1981a). The state of moult in each feather can be scored on a simple scale of 0–5 and the complete set of primaries on both wings scored from 0–100 (Ashmole 1962). Quantitative comparisons can thus be made of the speed of moult and completion dates and duration can also be estimated.

Because of the age differences and high degree of variation among individuals described above, such calculations are rarely possible for barn owls and the use of a numerical score to chart the progress of moult is not particularly instructive. A simpler approach, to gain some understanding of the timing, is to examine the percentage of birds that have shed or are growing new feathers each month and to quantify the number of feathers actually in growth each month. I did this over a period of 12 years in southern Scotland, spanning high and low vole years. It was thus possible to compare moult in male and female owls in relation to time of year, breeding, and yearly variations in food supply.

In general, females began moulting their wing feathers about 2 months earlier than did males. Trapped females were found in wing moult from the beginning of May to the end of October and males from July to November. Freshly moulted primaries and secondaries were found at the roost sites of a small number of untrapped females at the start of November. Peak primary and secondary growth among females occurred in July whereas, in males, August and September were the most active months (Figs. 8.3, 8.4).

When examined in relation to breeding (Fig. 8.5), about one quarter of females shed their first primaries and secondaries during incubation. The proportion increased steadily to reach a maximum in the second half of the nestling stage when about two-thirds of all females were in moult. There was clearly some interaction between calendar date, stage of breeding and moult, as none of the early laying females that completed their incubation before late April started moulting during incubation. Thus in 1981, for example, a peak vole year, none of 17 females that completed incubation in April began moulting in April, whereas 8 out of 24 females whose incubation was mostly in May, also started moulting in May. Also, females that laid in June did not start moulting until then. This is perhaps not unexpected as the two possible proximate factors controlling moulting times are day length and food supply (Voit-kevitch 1966, Stresemann and Stresemann 1966). The birds may not have been able to begin moult until a particular day length regime initiated hormone secretion and this may have been further modified by the balance of food intake rate and the energy requirements of laying. A much higher percentage of females began primary and secondary moult during incubation in high vole years than in low vole years, thus strengthening the argument that food supply was indeed important in the timing of the start of moult (Fig. 8.6).

In contrast to the females, most males did not begin wing moult until the young were well grown and peak activity occurred in August and September, after most of the young had become completely independent of parental care. This difference between the sexes makes considerable sense. There was a clear advantage for females to replace as many flight feathers as possible during incubation and the early nestling stage, when

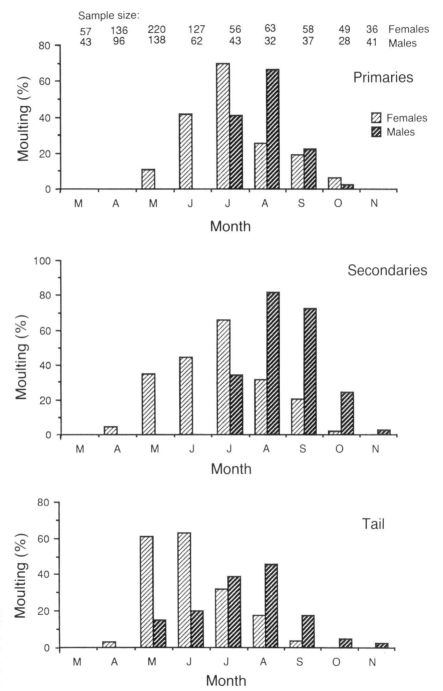

Fig. 8.3. The timing of wing and tail moult in breeding males and females, south Scotland. Males tended to moult later than females. Sample sizes shown along the top.

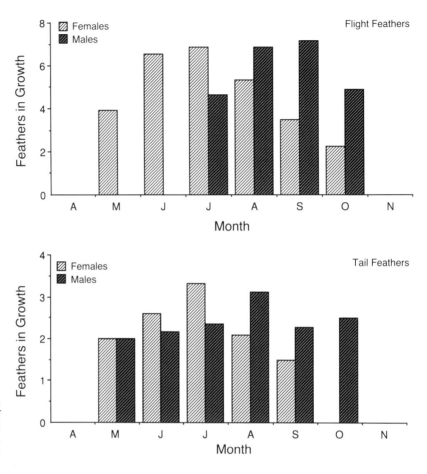

Fig. 8.4. Seasonal change in the average number of feathers in active growth in male and female barn owls, south Scotland.

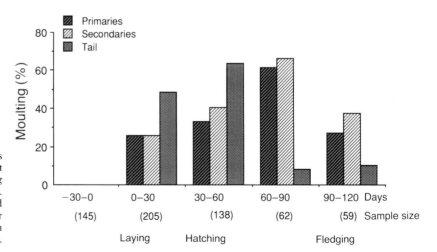

Fig. 8.5. Female barn owls undertake much of their moult while incubating and brooding small chicks (south Scotland). Graph shows days before and after laying of the first egg for each female (sample size given in parentheses).

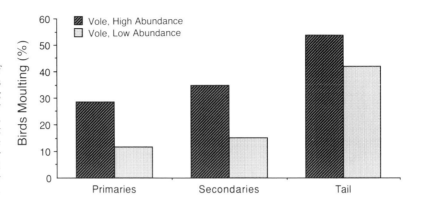

Fig. 8.6. The percentage of breeding female barn owls starting moult during incubation in high and low vole years, Esk study area, south Scotland. Wing moult starts later when prey is scarce (for primaries, $p < 0.01$; for secondaries, $p < 0.01$) but tail moult is not significantly later ($p > 0.05$).

they spent almost all their time at the nest. As they were not hunting, reduced flight efficiency would have been of no consequence for them. The males, not yet burdened with the demands of large, growing chicks could afford to devote some of their hunting effort to providing the females with the added food requirements for moult. There may also have been another, not so immediately obvious, advantage to the females. There has been no published account of the energetic requirements of moult in barn owls but this has been thoroughly researched in European kestrels (Masman 1986). When moult was at its most intense, the bird's basal metabolic rate (BMR) was increased by about 30%. Some of this arose from feather production but most was the result of an increase in heat loss arising from reduced insulation and greater vascularisation and blood flow to the skin. Long-eared owls experience a rise of 25.3% in BMR during the peak of moult (Wijnandts 1984). By undertaking a significant proportion of the moult whilst being largely inactive and in the insulated microclimate of a cavity nest, female barn owls were probably able to reduce substantially these added energy needs and thus make maximum advantage of the food supplied by the male.

It is interesting to consider the timing of the female's moult in relation to the theory of sexual conflict which proposes that a parent should minimise its share of parental care at the expense of its mate while ensuring that breeding success is not reduced (Trivers 1972). By 'using' the male to supply her with the nutrients she needs to complete a substantial part of her moult, the female barn owl can devote time to producing a second brood, either with the same mate or another mate (this has been recorded twice in the Scottish area), and still have completed moult before the start of winter.

Wing moult in males occurred mostly after the young had fledged, suggesting that either a reduction in flight efficiency or an increase in energy requirements during moult were incompatible with the energetic demands of raising a brood. Further evidence in support of such a conflict was provided by the behaviour of unmated males, males that reared two broods and males that reared exceptionally large broods. Unmated males were often difficult to catch as they tended to roost in open buildings and were wary of humans, but their moulted feathers could be found under their perches. Out of 23 such males located, 12 shed primary and secondary feathers in June whereas none of 62 breeding males examined were moulting at this time (chi-square = 37.63, $p < 0.001$). Only about 6% of males reared second broods in Scotland but those that did so were still feeding young in October. Freshly moulted secondaries and tail feathers were found at their roosts during November and early December, substantially later than in single-brooded males, suggesting that their moult had been delayed. It would be instructive to quantify this precisely but the question could better be tackled in continental Europe where second broods are much more frequent than in Scotland.

Some males that reared exceptionally large broods moulted very few of their flight feathers. In the peak vole year of 1981, two males that reared broods of seven and one that reared a brood of eight, failed to replace any of their primaries. These birds were all at least 4 years old and some of their primaries were in extremely poor condition with numerous breakages. All three were at very low body weights (between 260 and 280 g) by the time the last of their young finally left the nest area around the end of August. It seems likely that the need to recover condition took precedence over moulting. This again would merit further study but a major limitation here is the difficulty of catching the males after they stop feeding young.

Good evidence for the conflict between the energetic requirements of breeding and moult has been provided for Ural owls in Finland (Pietiainen, Saurola and Kolunen 1984). Failed breeders and non-breeders moulted more flight feathers than successful breeders and among the latter there was an inverse relationship between the number of feathers replaced and their brood size.

Moult in temperate region barn owls obviously occurs in the summer and autumn months when temperatures and prey abundance are relatively high. In humid, equatorial environments, temperatures vary little and small mammal populations tend to show greater seasonal stability. Nevertheless, barn owls still demonstrate a marked seasonality in their moult. In Malaysia, Lenton (1984b) inspected 11 nest sites regularly each month over 2 years for moulted feathers. He was unable to distin-

guish male from female feathers but it is likely that the sample was
biased towards females. Freshly shed primaries were found in almost
all months but there was a marked peak each year between March and
July, coinciding with the seasonal low in breeding activity in the owl
population. Two captive pairs completed most of their moult before
starting to lay their eggs.

It seems that in this tropical population there was an even more
marked separation of moult and breeding than has been found in temperate region populations. The reason for this probably lies in the different seasonal temperature regimes. Masman and Daan (1987) showed
that the extra energy required for moult by European kestrels would be
considerably increased were they to undertake the moult in winter,
mainly because of the enhanced rate of heat loss at lower temperatures.
The same probably holds for barn owls. In northern areas, barn owls
often have difficulty meeting their normal energy requirements in winter,
without having to add to it with moult. They therefore probably have
no option but to have a greater overlap of moult with breeding and to
end moult before the onset of low winter temperatures. Tropical barn
owls have no such temperature limitation and this may be the explanation
for their apparent ability to achieve a greater separation of moult and
breeding.

8.3 Evolutionary aspects of moult

Temperate region barn owls show some similarities to and some marked
differences from other temperate birds of prey. The males of European
kestrels (Village 1990) and sparrowhawks (Newton and Marquiss 1984)
also tend to moult their flight feathers later than do females. In both
species, females start during incubation and males some 20–30 days
later. Female barn owls thus show much the same timing as female
kestrels and sparrowhawks but male barn owls clearly undertake most
of their moult much later than the males of these species. Kestrels and
sparrowhawks also tend to replace all their flight feathers each year,
whereas temperate region barn owls take 2 to 3 years to achieve this.

Are these differences related to differences in the ecology of the
species? The observation that tropical barn owls replace all their flight
feathers each year in the same way as kestrels and sparrowhawks do
suggests that there might be a significant disadvantage in continuing
with old feathers. Temperate area barn owls not infrequently are seen
to carry badly damaged and worn flight feathers from one year to the
next. Broken and worn feathers would be expected to impair flight
efficiency but also they might reduce the noise suppressing qualities of

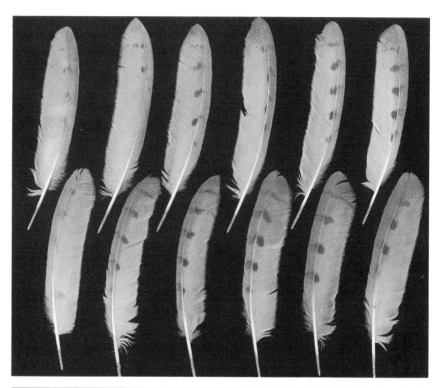

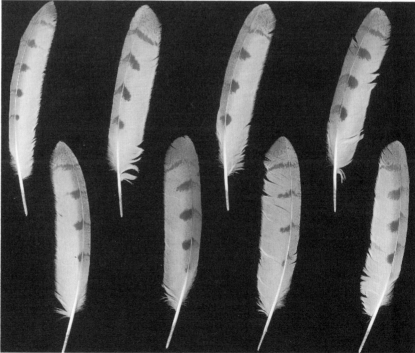

Fig. 8.7. Moulted primary feathers can be a useful aid to the identification of individual birds. Upper plate shows a number of different males, lower plate, a number of females. The pattern is unique to each bird. (Photograph: Iain R. Taylor and Dave Haswell.)

the wing, thereby possibly lowering hunting success. The postponement in the replacement of these feathers suggests that strong selective forces are operating against a complete annual moult or in favour of other activities whose energetic or other demands preclude a complete moult.

Evidence for a conflict between the requirements of breeding and moult have been discussed earlier. A major difference between the barn owl and species such as the kestrel, with which it often shows some similarity in diet and habitat in Europe, is the number of young reared. Barn owls commonly produce four to six young and sometimes up to eight or nine in a single brood, whereas kestrels normally rear only two to four young. Barn owls will also rear second broods when food conditions permit whereas kestrels apparently will not (Village 1990).

The moult pattern of temperate zone barn owls might therefore be a strategy to enable maximum productivity whilst minimising the risk to survival that a complete moult might entail in an environment in which prey abundance shows marked fluctuations and where there is a cold winter season. The severely restricted moult shown by first year breeders, involving most frequently only one of the primaries and a small number of secondaries and tail feathers, may represent the extreme of such an evolutionary strategy. These inexperienced birds breed later and less successfully than older birds (Chapters 9 and 12) and presumably are under greater stress during breeding. Also, as many of these birds will survive to breed only once, it may be more important for them to maximise their breeding output than to improve subsequent survival by the replacement of feathers.

Thus the moult pattern may be an integral part of the whole complex of adaptations shown by barn owls that enables them to exploit temporary and unpredictable food supplies.

8.4 Summary

Barn owls in temperate areas normally moult their juvenile set of flight and tail feathers over three moult phases, starting in their second year and ending by their fourth year. Thereafter, replacement occurs over a 2-year cycle. Birds in a tropical population studied in Malaysia moulted all flight and tail feathers each year starting with the replacement of juvenile feathers towards the end of their first year. Primaries are replaced ascendantly and descendantly from a centre at P6. Secondary moult usually begins at S12 and tail moult at the central feathers (T1).

Female barn owls in Scotland start their annual moult in May, reach a peak of growth in July and are usually completed by the end of October. Males may begin tail moult in May, but wing moult is delayed until July and usually completed by early November. Many females therefore start to moult whilst incubating whereas males do not begin until the young are almost fledged. Moult is influenced by yearly variations in food supply: significantly fewer females begin moult during incubation in low vole years compared with high vole years.

Several aspects of moult in temperate area barn owls, such as the slow pattern of replacement which, in firstyear birds, most often only involves one of the primaries, and the very late timing in males, suggest an evolutionary strategy which minimises the conflicting energy demands of moult and breeding and gives priority to the production of offspring. This may be an integral part of an overall life history pattern that enables barn owls to exploit temporarily abundant food supplies.

9 Breeding seasons

The timing of breeding in birds, including owls, has presumably been subject to natural selection just as much as anatomical or behavioural traits. Generally, selection will favour those individuals that breed at a time that results in most offspring being raised and surviving to be recruited to the breeding population in subsequent years. Many owls are relatively long-lived, so one might expect breeding seasons to be shaped by both the production of young and the survival of the adults such that the bird's lifetime reproductive output is maximised. Thus the timing of breeding should be considered in relation to other aspects of the bird's annual cycle that might influence survival such as the need to replace worn feathers or to recover body condition before the onset of less favourable conditions.

When considering the ecological and evolutionary significance of breeding seasons, it is useful to distinguish between ultimate and proximate factors. Birds usually breed when food availability and weather conditions that determine energy requirements are such that the female can form eggs and rear young. Thus food supply is generally considered to be the most important ultimate factor controlling breeding seasons. For the barn owl, changes in food supply can occur through changes in prey abundance, or through changes in weather and vegetation that affect small mammal activity and accessibility.

If birds are to breed at the most appropriate time they must either possess endogenous rhythms that initiate breeding or they must be able to respond to environmental cues. Such cues, or proximate factors, could in theory be any aspect of the environment that gives a reasonably good prediction of subsequent conditions. Many birds, particularly temperate species, respond physiologically to changes in day length: they are said to be photoreceptive. Following a period of short days in winter, increasing day length triggers the secretion of hormones from the pituitary gland that stimulate the initial development of the reproductive organs. Different species are sensitive to particular characteristic levels of day length (critical day length) and so may begin breeding at different dates, appropriate to their requirements. Responsiveness to changing day length may provide the bird with initial information and prevent breeding at inappropriate times of year, but may not be enough on its own to bring about laying. Additional stimulus is needed and it is likely that in most species this is provided by the level of food supply experienced by the female.

Perrins (1970) suggested that females start breeding as soon as they have accumulated enough reserves of energy and nutrients to form eggs. This suggestion probably has to be modified slightly for owls and other birds of prey as there is evidence that the female must also build up enough reserves for use during incubation to act as an insurance against temporary food shortage (this study, Hirons 1976, Newton 1979). In support of this idea, laying dates of European kestrels (Cavé 1968, Dijkstra et al. 1982), Tengmalm's owls (Kopimäki 1989) and sparrowhawks (Newton and Marquiss 1981b) have been advanced by the provision of supplementary food early in the breeding season. To my knowledge no such experiments have been conducted with barn owls but it is likely they would respond in the same way.

The above argument probably applies to most birds living in temperate climates. The proximate cues used by birds in the tropics are much less clearly understood. For those close to the equator, day length changes are negligible and probably cannot be used. Other factors, such as rainfall, temperature or changes in food supply directly, have been suggested. The same might apply to opportunistic breeders that can breed at any season depending upon irregular and unpredictable food sources as might be the case for barn owls in parts of Australia or other more arid parts of their range.

The cosmopolitan distribution of the barn owl presents a unique opportunity to compare breeding seasons in different parts of the world and to investigate ultimate and proximate causes in different environments.

9.1 Geographic variation in breeding seasons

Barn owl laying dates have been studied extensively in Europe and North America but rather less so elsewhere. In Fig. 9.1, I have plotted the findings from 11 studies spanning most of the latitude range over which barn owls occur and a more detailed comparison is made in Fig. 9.2 of a temperate population in Scotland with a population in France and a tropical population in Mali in West Africa.

All populations that have been studied, including those within the tropics, show at least some seasonality of breeding. Close to the equator, new clutches have been found in almost every month, but laying still tends to be concentrated within specific periods. Moving away from the equator, seasonality becomes more distinct with definite periods of non-breeding.

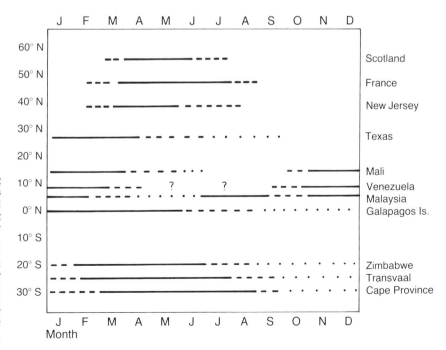

Fig. 9.1. The spread of laying from 11 studies of barn owls at different latitudes. Solid lines show main laying seasons, broken lines show extremes of laying periods. Sources, from top to bottom: Taylor, this study; Baudvin 1986; Colvin 1984; Otteni *et al.* 1972; Wilson *et al.* 1986; Lander *et al.* 1991; Lenton 1984b; de Groot 1983; Fry *et al.* 1988; Fry *et al.* 1988; Fry *et al.* 1988.

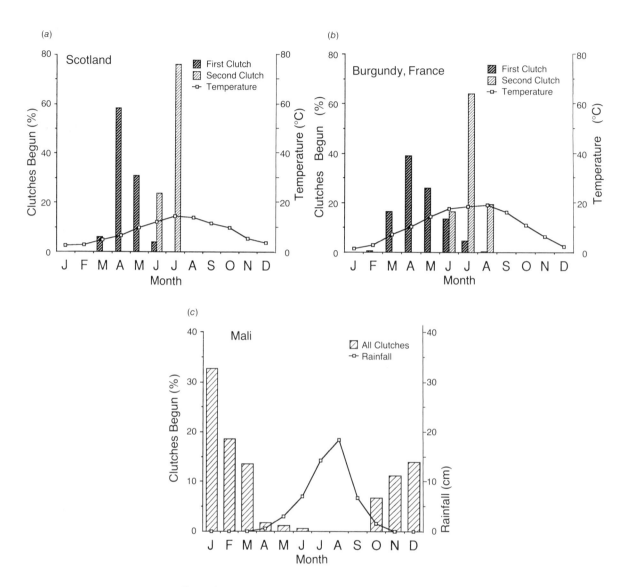

Fig. 9.2. Comparison of laying dates in Scotland, France and Mali. In savannah habitats, breeding is related to rainfall; in temperate areas it is related to temperatures. (Sources: see Fig. 9.1.)

Breeding in northern temperate regions coincides with the summer increase in temperature and hence the period of annual biological productivity. However, laying starts early, usually before or just about the time their small mammal prey start to breed. Thus laying begins whilst food abundance is still declining (Chapter 5) and a significant increase in food supply does not occur until about 2 or 3 months later when the young small mammals become independent and large enough to catch, by which time most owls have half-grown young in the nest. The onset of breeding cannot therefore be brought about by an increase in prey abundance and it is likely that the most important factor is the general amelioration in weather conditions. This would tend to increase the activity levels of small mammals and hence increase their availability to the owl, and also reduce the owl's maintenance requirements. This would result in an increase in net energy and nutrient gain by the females and would be aided by a decreasing frequency of the adverse weather conditions such as sleet and heavy rain that inhibit the owl's foraging activity (Chapter 6). The increase in territorial and general sexual activity among the prey in early spring must also substantially increase their availability to the owls and probably has a significant influence on laying dates (Chapter 5).

It is an advantage for barn owls in temperate regions to start breeding as early as possible because the ability to rear a second brood depends greatly on the timing of the first brood (Chapters 11 and 12). Also, by laying mostly in April and May, young from the first brood leave the nest and become independent from about July onwards, when small mammal numbers are at a high and still increasing density and when weather conditions remain favourable. Thus they can learn to hunt effectively before the onset of winter. Early-fledged young have a higher probability of surviving to be recruited to the breeding population than later fledged young (Chapter 15). Early laying also gives the adults a period of high prey abundance and reasonably high temperatures during which they can meet the energy demands of moult.

Barn owls in Britain rarely produce clutches after July, even though prey abundance is higher then than in spring. It takes almost 4 months to produce an averaged-sized brood, so eggs laid in July would result in the young becoming independent in November, just before the onset of winter conditions. One would expect such young to suffer a high mortality and indeed late broods in Scotland have low success even before they leave the nest. In addition to this, the adults, especially the males, would have little time to complete their moult before winter. As a result, in most years, natural selection would be expected to operate against late breeding.

The seasonal pattern of laying, especially in British barn owls, is not symmetrically distributed around mid-summer but is strongly skewed towards the spring. This type of pattern is found commonly among species in which the termination of breeding is controlled by day length. Such species become insensitive to the stimulatory effect of long days after prolonged exposure and thus breeding is switched off when day lengths are still greater than those that stimulated breeding in the spring (Lofts and Murton 1968). The responsiveness of barn owls to changing day length has not been tested; we do not know if this is responsible for switching off breeding or, for that matter, for initiating breeding in the spring.

An interesting comparison can be drawn between British barn owls that depend mainly on the field vole and those of continental Europe that depend on the common vole. Laying of first clutches tends to end slightly earlier in Britain and fewer second clutches are laid (Fig. 9.2). Field voles are found only in long grassland whereas common voles can occur widely in much shorter vegetation such as grazed pastures and in crops. Thus it is possible that, in Britain, the seasonal increase in vole abundance is progressively offset by the growth of grasses, thereby limiting the improvement in prey availability and curtailing the breeding season. Continental barn owls may experience increased prey availability for much longer and, together with more favourable weather conditions, may allow a more extended breeding season. Thus it is possible that the timing of laying in all European barn owls is controlled directly by food supply with no influence of day length.

Tropical and sub-tropical regions experience high temperatures throughout the year and rainfall becomes a more significant determinant of seasonal productivity. Throughout Africa, barn owls seem to lay almost entirely during the dry seasons. In Mali, some pairs lay immediately after the cessation of rains but the majority do not lay until the middle of the cool (relatively) dry season some 3 to 4 months later (Fig. 9.2, Wilson *et al.* 1986). Obviously, rainfall cannot be the proximate stimulus to breeding. Many small mammals such as the multimammate rat, one of the most important barn owl prey species in Africa, have breeding seasons extending from the latter half of the wet season into the early dry season, with numbers declining steadily thereafter (Chapter 5). African barn owls thus start laying on a high prey density and complete their breeding on declining prey densities, a markedly different situation from that experienced by temperate zone barn owls. However, once again, seasonal changes in vegetation must be considered. Growth during the wet season is prodigious, especially in grassland areas, and access of owls to small mammal prey must be dramatically reduced, increasing only as the vegetation begins to die back or as fires remove the annual accumulation. Increasing prey availability, brought about mainly by seasonal changes in vegetation, may be responsible for the timing of breeding in African populations and may indeed be the most important proximate factor in other tropical areas with very marked wet and dry seasons. For example, populations in Venezuela also lay during the dry season (Lander *et al.* 1991). In these populations then, food supply would act as both the ultimate and the most important proximate factor.

Rather less information is available for arid tropical areas where seasonal changes in vegetation are unlikely to be so important. Galapagos barn owls, which occupy the relatively dry parts of the islands, can apparently lay at almost any season but most do so during the wetter part of the year. This corresponds with the breeding season of rodents and the period of increased activity of insects that form a small but perhaps important component of the diet there (de Groot 1983). As the Galapagos Islands straddle the equator, day length changes are minimal and it is unlikely that the birds can use light conditions as a proximate cue. This is supported by the observation of birds with clutches in almost all months. Probably, these birds respond directly to an increase in prey availability brought about by an increase in prey abundance. Laying also occurs in Texas during the wetter part of the year, again coinciding with the breeding of small mammals (Otteni *et al.* 1972).

No quantitative data seem to be available for Australia, but anecdotal reports suggest a dynamic system with large-scale movements and irregular breeding associated with temporary increases in small mammal populations (Fleay 1968, Morton *et al.* 1977, Morton and Martin 1979, Hollands 1991).

9.2 Variation in laying dates within populations

All barn owl populations with regular, distinct breeding seasons show annual variations both in the start of laying and in mean laying dates. These two were significantly correlated in the Scottish population ($r = 0.93$), with mean dates of the start of the first clutches varying from 9 April to 26 May, a range of 47 days over 12 years. In the French populations studies by Hugues Baudvin (1986) mean dates ranged from 16 April to 8 June, a 53-day interval over 8 years. The New Jersey population was more consistent from year-to-year, with only about 15 days between the earliest and latest seasons over five years (Colvin 1984).

Studies of other species have shown correlations between laying dates and spring prey abundance (e.g. Tengmalm's owls, Kopimäki 1986), or temperatures and rainfall immediately before laying (e.g. blue tits, Perrins 1970; sparrowhawks, Newton and Marquiss 1984).

To investigate these relationship in the Scottish barn owls, I performed a stepwise multiple regression analysis using data from 12 consecutive years (Table 9.1). The variables entered were spring prey abundance and weather conditions (temperatures and rainfall) and winter weather conditions prior to breeding (temperatures, rainfall and snow cover). Spring prey abundance and temperature varied independently

Table 9.1. *Laying dates of individual females in relation to field vole abundance and various aspects of winter and spring weather. Results of a stepwise regression analysis in the Esk study area, south Scotland*

Variable	Partition coefficient	F	p
Field vole abundance	−0.504	86.12	< 0.001
Average March–April temperatures	−0.26	26.80	< 0.001
Average December–March temperatures	−0.067	0.47	n.s.
Days with snow lying December–March	0.041	0.19	n.s.
Wet days March–April	0.067	1.62	n.s.
Wet days December–March	0.188	11.7	0.01

n.s. not significant.

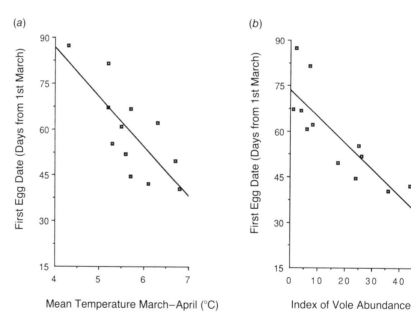

Fig. 9.3. Correlation between the mean dates of laying of first eggs each year over the whole of the Esk study population and (*a*) average March–April temperatures, *r* = −0.72, *p* < 0.01 (*b*) spring vole abundance, *r* = −0.81, *p* < 0.01. Together, these two variables explained 86% of the annual variation in laying dates (on the *y*-axis, 1 is 1 March).

of each other. The analysis showed that mean laying dates (of first eggs) were closely correlated with both spring prey densities and temperatures immediately before breeding (Fig. 9.3). The owls laid earliest when prey abundance and spring temperatures were high. Of these two, prey abundance was the more important and together they accounted for 86% of the total annual variation in mean laying dates.

When mean laying dates are plotted against year and alongside the spring index of field vole abundance, they can be seen to follow a

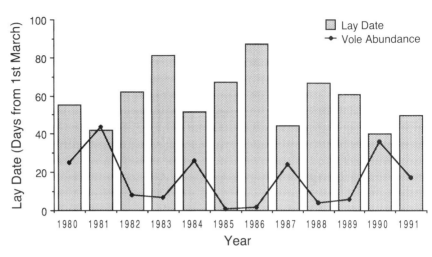

Fig. 9.4. The mean dates of laying of first eggs each year follows a distinctly cyclic pattern correlated with the vole cycle, Esk study area, south Scotland (on the *y*-axis, 1 is 1 March).

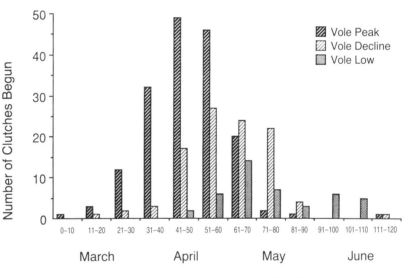

Fig. 9.5. The spread of laying of first eggs during the 3-year vole cycle: peak, decline and low vole springs, Esk study area, south Scotland (1 is 1 March).

consistent cyclic pattern, perfectly in synchrony with the 3-year vole cycle (Fig. 9.4). Laying is earliest in peak vole springs, is later in the year of decline and later still the following year (Fig. 9.5). Later laying in the low vole years could be explained simply on the basis of food supply but this does not explain why the third year in the cycle should be so much later than the second year, when vole densities in the spring were not markedly different. In peak years, the owls not only had high spring food supplies but also had experienced high prey densities over winter. In the decline year, vole abundance was low in spring but had been relatively high over much of the winter. Springs in the last year of the cycle also had low vole abundances but in these year densities were low over the whole winter (Chapter 5). Thus it is possible that in the third year the birds started the spring at lower body weights and that they took longer to build up the reserves needed for breeding. Unfortunately too few data on female weights were collected during the early springs of these years to test this idea thoroughly. Cavé (1968) showed that oocyte development in European kestrels is initiated in the autumn and proceeds throughout the winter and early spring. Winter food levels affected the rate of growth of the oocytes. It is possible that a similar effect operates in barn owls.

The correlation between laying and spring temperatures could have come about either directly by affecting the bird's energy requirements or indirectly by altering the activity of small mammals and hence the bird's energy or nutrient intake rates. Further evidence of the importance of spring temperatures was provided by a number of cases when there was a recurrence of exceptionally cold conditions during laying. Periodic cold snaps often occurred through to early May and in 1981, for example, temperatures were very low between 24 and 28 April. Three females who were in the process of laying gave up temporarily, laying their next eggs 6 to 9 days later rather than at the usual 2-day interval. Nonetheless, all their eggs hatched.

Spring rainfall had no influence on laying dates and nor did winter temperatures or snow cover. No relationships were found with temperatures or snow when December to March values were substituted with values for January to March or February to March. Severe winters were therefore not associated with later laying at least not within the range of winters experienced during the study. There was some evidence that high winter rainfall was associated with late laying but, although statistically significant, this relationship was weak and may not have been real. It might be expected that rainfall would impair the owl's hunting efficiency (Chapter 6). In winter, heavy rain is often associated with strong and blustery winds that generates a great deal of noise from vegetation which might make it harder for the owls to detect prey. Also rainfall at low winter temperatures might reduce small mammal activity (Chapter 5). The reason for a lack of relationship with spring rainfall may be that absolute amounts of rainfall in spring, especially in April, were much less than in winter months and temperatures were much higher; thus, it perhaps exerts less of an influence on prey activity.

The finding that laying dates in Scotland were not correlated with temperatures in the previous winter contrasts with the situation in France where Baudvin (1986) reported a good correlation for the years

1972 to 1979. However, in an immediately adjacent study area, Chanson *et al.* (1988) demonstrated no relationship between winter temperatures or amount of snow and laying dates between 1977 and 1986. Both studies included a wide range of winter conditions from very mild to exceptionally cold so the contradictory results cannot be attributed to differences in the sample periods. In neither case were the authors able to assess the abundance of small mammals in spring so a complete analysis of the situation is not possible until clarified by further evidence. Continental areas can experience more prolonged periods of severely cold conditions than the more temperate coastal parts of Europe and some small mammals, such as the common vole, may be adversely affected, especially as this species is not always able to forage under an insulating layer of long grassland (Chapter 5).

9.3 Effect of habitat on laying dates

Local variations in habitat occur in most areas inhabited by barn owls. In the Scottish area, pairs nested in young conifer plantations, upland grasslands and lowland pastoral farmland and there were consistent differences in the laying dates of birds nesting in these habitat types (Fig. 9.6). Those in plantations laid earliest and those in the low farmland latest. The difference was marked, with an average of 14 days between them. Plantations provided extensive areas of prime field vole habitat whereas in the low farms vole habitat was largely restricted to

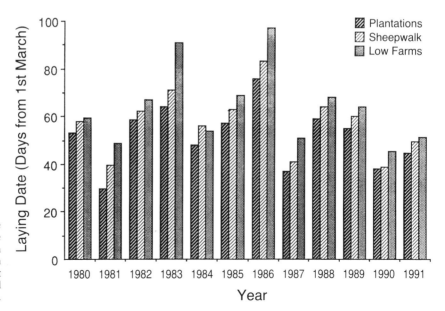

Fig. 9.6. Average laying date in relation to habitat in the Esk study area, south Scotland. Pairs nesting in plantations tend to lay earliest and those in low farmland habitats lay latest.

field edges. Therefore the laying date differences were probably a result of different levels of prey abundance around the nest sites.

The average altitude of plantation and low farm nests was 209 m and 89 m respectively. As a result, plantations pairs experienced considerably more severe winter conditions, with lower temperatures, higher rainfall and more days of snow cover (Chapter 1). The much earlier laying dates of these birds, despite these weather differences, further emphasises the unimportance of winter weather to laying dates.

Habitat variations in laying dates, although not as great as those above, have been noted in New Jersey by Bruce Colvin (1984) and in France by Hugues Baudvin (1986). Again, those pairs nesting in prey-rich habitats laid earlier.

9.4 Effect of female size and age on laying dates

Within any year there was a great spread of laying dates within the Scottish population and the same is true for most other populations that have been studied in detail (Fig. 9.6). Some of this variation was the result of pairs nesting in different quality habitats, as discussed above, but much variation still remains to be accounted for.

No relationship was found between laying date and female wing length. However there was a strong effect of age, with first year females laying much later than older females (Table 9.2).

Almost nothing is known of the pattern of establishment or timing of courtship in first year birds. I have trapped some that had settled at their future breeding site within 2 or 3 months of fledging, but many were not present until a few weeks before breeding. Some established females died and created vacancies as late as the end of March, so some first year females presumably settled and started courtship relatively late. It is also possible that young females, being less experienced, took longer to build up body reserves before breeding. The average weight of females increased from 350 g about 40 days before laying to a peak of 425 g at laying (Fig. 11.1, p. 150). By parting the breast feathers, it could be seen that part of this weight increase was achieved by the accumulation of subcutaneous fat, but the pectoral muscles also increase in size, as gauged against the sternum. These reserves, especially the proteins in the muscle mass, may be needed to produce the clutch but this cannot be their only role as weight loss occurs progressively through incubation. Females that abandon their clutch during incubation are significantly lighter than successful females (Fig. 11.1, p. 150), and it is likely that the body reserves serve as an insurance against the possibility of poor

Table 9.2. *Comparison of the date of laying of first eggs by first year females and older females, Esk study area, south Scotland*

Year	Laying date[a]		t-difference	p
	First-year females	Second-year or older females		
1981	54.4 ± 3.4 (13)	37.2 ± 2.2 (28)	4.31	< 0.001
1982	76.5 ± 2.6 (8)	57.3 ± 2.3 (24)	4.53	< 0.001
1984	57.1 ± 2.1 (9)	47.5 ± 1.9 (17)	3.20	0.004
1987	51.0 ± 2.9 (7)	40.9 ± 2.5 (13)	2.55	0.02
1990	52.5 ± 9.4 (8)	37.8 ± 2.3 (17)	2.07	0.05
1991	50.9 ± 4.9 (10)	43.5 ± 2.4 (15)	1.50	0.15

[a]Laying date takes day 1 as the 1st March and is expressed as a mean ± S.E.M. with the number of observations in parentheses.

Table 9.3. *Assortative mating of first year and older birds, south Scotland. The figures are from high vole years only, when substantial numbers of young birds were recruited to the breeding population. First year males tend to pair up with first year females (p < 0.005)*

	Male	
	Adult	First year
Adult female	53	17
First year female	21	24

hunting conditions occurring during incubation. A similar argument has been advanced for tawny owls, sparrowhawks and European kestrels which show similar changes in body weights (Hirons 1976, 1985, Newton 1986, Village 1990). However, male experience is probably more important than female experience in the accumulation of these reserves. Males start presenting food to the females at least a month before laying and this increases progressively so that for the last 2 weeks or so they are responsible for most, and perhaps even all, of the female's intake. First year females breed more frequently with first year males than with older males (Table 9.3). This is probably not an active selection process but merely a consequence of temporal changes in recruitment patterns.

Large numbers of first year birds enter the breeding population in peak vole years (Chapter 15), but the survival of older birds is also highest in these years (Chapter 14). Therefore, most first year birds occupy previously vacant sites and, since older birds are highly faithful to their sites (Chapter 13), the first years tend to end up breeding with mates of the same age.

Thus, later laying in first year females might be more a result of the age and lack of experience of their mates rather than their own age. This could be explored further by comparing pairs where first year birds are mated with older birds, but few such pairs were recorded in any one year and combining years was not useful because of the large annual variation in laying date.

9.5 Summary

Barn owls nesting close to the equator can lay eggs in almost any month but some seasonality is still evident. Moving north and south, laying seasons become more distinct with definite periods when no laying occurs. In some hot arid areas, such as parts of Australia, breeding may be sporadic and irregular.

In tropical or sub-tropical grassland habitats, where there are marked wet and dry seasons, barn owls usually lay well into the dry season. As their prey populations reach high levels earlier, towards the end of the wet season, it is likely that the annual die-back of vegetation through its effect on prey availability is a more important determinant of breeding seasons than prey abundance alone. In arid tropical areas, where vegetation changes are less important, breeding may occur during wet periods, coinciding with rodent breeding activity. Laying in temperate regions occurs mostly from March to June, coinciding with the onset of higher temperatures. In these regions, young leave the nest in mid to late summer when prey populations are increasing; the young can thus gain experience before the onset of harsh winter conditions.

In the Scottish area, yearly variations in mean laying date were correlated with spring vole abundance and spring temperatures, but not with winter temperatures or snow cover. Laying dates followed a cyclic pattern corresponding precisely with the 3-year vole cycle with earlier laying in the peak vole years. Laying dates were also related to habitat, with pairs in conifer plantations laying earliest followed by those in sheepwalk and low farmland, correlating with the density of prey-rich foraging habitats around the nest sites.

No relationship was found between female size and laying date. First year females laid later, on average, than older females. There was assortative mating with first year females pairing more often with first year males than expected by chance. Male experience, by assisting the female to accumulate body reserves through courtship feeding before laying may be a more important determinant of laying dates than female experience.

10 *Nest sites*

Barn owls do not not construct a nest in the sense that they do not deliberately carry in nest materials. However, once a site has been chosen, the birds start by making a slight scrape or depression and, during the course of incubation, the female pulls in the fur from the pellets she casts around her so that eventually the eggs come to sit on a soft and presumably insulating pad of matted small mammal hair. Barn owls are cavity nesters and in their natural state must have been totally dependent on tree holes, cracks and fissures in cliffs and the old nests of other species. In parts of the world that still retain large areas of natural or semi-natural habitat, such as much of Africa and Australia, most barn owls probably still nest in this traditional way. However, throughout most of North America and Europe, a high percentage of barn owls now nest in association with humans, making use of a vast variety of artificial structures, a characteristic that has given rise to some of the species' common names such as barn owl and kerkuil (church owl). This habit makes the bird highly vulnerable to changes in building fashions, especially of agricultural buildings, and can be a significant factor in the conservation of the species, a subject which is examined later (Chapter 16).

The ease with which barn owl nests can be found varies greatly according to their nature. Nests in tree holes are generally much more difficult to track down than nests in buildings and, among buildings, those in conspicuous structures are the most readily located. Therefore in France or Germany, for example, it is relatively straightforward to visit the church towers to check for the presence of a nest but to search all other possibilities in farms and villages would be extremely difficult and time consuming. Researchers have rarely attempted to locate all nest sites within given areas and it is likely that most published accounts of nest sites are strongly biased in favour of the more easily located types. Therefore caution must be exercised especially when comparing different areas or when examining historical changes. The situation can be further complicated when nest boxes are provided.

It is important to have a clear idea of why we should be interested in nest sites in the first place. There are several answers to this. We might want to discover if the frequency of utilization or the breeding success is related to nest type. We might want to find out if barn owl populations are limited by the availability of nest sites or if declines in population are caused by diminishing nest site abundance. We might want to know if there is any competition for nest sites with other species. It will usually be important to answer these questions in areas that are typical of the population as a whole, in other words not in areas that are the subject of specific conservation programmes that alter nest site type and availability by the provision of artificial nest boxes. The examination of 'natural' nest sites might however help in the design and location of such nest boxes, for example by quantifying preferred heights or dimensions.

10.1 Types of nest site used

Nests in trees, cliffs and other bird nests

Despite the obvious difficulties in identifying sites in trees and the biases that must be present in much published work, it is clear that there are very great geographic variations in the use made of these nesting places. In his New Jersey study area, which was searched very thoroughly for nests, Bruce Colvin (1984) found 50.7% of 209 breeding attempts in tree holes. When nest boxes were excluded, this value rose to 73.6%.

Fig. 10.1. Tree cavities must originally have been one of the main nesting places for barn owls. Left, an ash in south Scotland; right, a silver maple on a New Jersey farmstead – the cavity is about 5 m above the figure. (Photographs: Iain R. Taylor.)

The most frequently used species were silver maple (Fig. 10.1), *Acer saccharinum* (66.7% of total trees used), sycamore, *Platanus occidentalis* (28.4%), and American basswood, *Tilia americana* (4.9%). Silver maple is often grown as an ornamental tree around farmsteads and houses and is particularly prone to forming cavities.

Studies in continental Europe have generally shown a very low percentages of nests in tree holes (Kaus 1977, Braaksma 1980, Juillard and Beuret 1983, Baudvin 1986), but, in England, hollow trees have long been important, providing about 30 to 40% of nest sites (Blaker 1933, Sharrock 1976, Bunn *et al.* 1982, Shawyer 1987). Large old English elms (*Ulmus procera*) and oaks (*Quercus* spp.), both of which frequently form deep, watertight hollows, have traditionally been the most frequently used, but the former is declining in significance as Dutch elm disease takes its toll. Ash (*Fraxinus excelsior*) and willow (*Salix* spp.) are also commonly used.

In three of my Scottish study areas (Esk, Dee and Nith) where I conducted intensive searches each year to locate all nests; tree hole sites were uncommon making up only 9.9% of 101 sites used during 14 years. In a fourth very small area around my home, 14 (73.7%) of 19 sites were in trees.

Elsewhere, barn owls have been recorded using tall eucalyptus trees in Australia, figs in Asia and baobabs in Africa. Palms are used throughout the tropics, and in rainforest areas that have been cleared for agriculture they often nest in the hollow decaying stumps of former giant forest trees (Taylor, personal observation).

There is great variation in the heights of the tree holes used by barn owls. In Britain this ranges from just above ground level to over 10 m with a modal height of between 3.5 to 6.0 m (Glue, in Bunn *et al.* 1982) but, in New Jersey, heights vary from 2.5 to 14.3 m with a mean of 6.8 m (Colvin 1984). The north American birds clearly use much higher cavities but, as we have no information on the relative abundance of holes at various heights, it is impossible to decide whether this difference reflects selection or just availability. However, it is possible that they take higher and less accessible holes to minimise the attentions of racoons which can be significant nest predators (Bruce Colvin, personal communication). There are no important nest predators in any part of Britain, except possibly for feral cats, so there is little disadvantage to birds nesting at lower heights.

British and North American barn owls tend to select isolated trees around farmsteads and villages and along hedgerows and woodland edges, rather than in the centres of woodland. Generally, cavities must be large and dry, and in many cases the trees selected are completely hollow with the nest platform being as much as 5 to 6 m below the entrance. In a British sample of 140 trees, Glue (in Bunn *et al.* 1982) found that 50% of nests were in a cleft or cavity in the main trunk and 19.3% were vertical 'chimney' chambers in snapped-off trunks or decaying stumps. A further 15% were in the cavity or branch clusters of pollarded trees and only 10% were in large hollow side branches.

Given that brood sizes not infrequently go up to eight or nine, barn owls have probably evolved a selection for very large holes in comparison with many other species. Tawny owls, for example, have brood sizes averaging only about two to four, and have evolved in woodland where the predation risk from tree climbing martens is probably high. They select small cavities which presumably can be better defended and are adequate for the number of young reared. Competition with barn owls for cavities is most unlikely except perhaps where there is an extreme shortage of smaller holes. In any such competition it is far from certain that the tawny owl would win in any case for although they are much heavier than barn owls, the latter are far more agile and their long legs and talons would provide formidable opposition. Barn owls nest later than tawny owls but they still visit their nest sites regularly during the winter and presumably defend them as well.

There is almost no quantitative information on the use of sites in cliffs and rock faces, although presumably the habit must be very widespread. Records are most frequent from the more arid parts of the barn owl's range. For example, de Groot (1983) recorded nests in the Galapagos Islands in cavities in the ground and in the walls of lava tubes and small volcanic cones. Along the sea coast cliffs of western Scotland they often use rock fissures and sometimes also the dense growths of ivy that commonly grow there.

One of the most interesting of nest sites is to be found in the savannah areas of Africa. Here barn owls regularly breed in old nests of the hammerkop, *Scopus umbretta* (Wilson *et al.* 1986, Fry *et al.* 1988). This somewhat unusual member of the heron family builds large, domed nests measuring about 90 cm × 120 cm and placed on the thicker branches of trees, often close to or overhanging watercourses. The structure, made of sticks, grass stems and other vegetation, is securely bound together with mud or dung and a narrow mud-lined tunnel leads about 50–60 cm into the nest chamber. The building style has presumably evolved to reduce predation and provide cover against torrential tropical storms. It is not surprising, then, that they prove so popular with the owls. In the same savannahs, barn owls have also been recorded breeding in the large communal nests of the sociable weaver, *Philetairus socius* (Fry *et al.* 1988).

Nests in buildings and other man-made structures

The range of sites recorded in this category is enormous, from church towers, castles and chateaux, farm lofts, haybarns and silos to watertanks, cavities in bridges and even down wells (Figs. 10.2–10.5). Throughout most of continental Europe, churches are often the most frequently recorded nesting places (Fig. 10.3) but, as these have usually been subject to selected searches at the expense of other potential sites, it is not possible to say whether or not they are always the most important sites, although this seems likely (Honer 1963, Schonfeld and Girbig 1975, Braaksma and de Bruijn 1976, Juillard and Beuret 1983, Pikula *et al.* 1984, Baudvin 1986).

Fig. 10.2. Castles make excellent nest sites but present problems for researchers! Upper: the Hermitage Castle, south Scotland; lower: the Chateau Porrentruy, Switzerland. (Photographs: Iain R. Taylor.)

Fig. 10.3. Throughout much of continental Europe, barn owls commonly nest in church bell towers. Near Dijon, France. (Photograph: Iain R. Taylor.)

Fig. 10.4. Mosques also provide good nesting places. In this case, the birds used the small front dome, gaining access through a small gap underneath. Near Selangor, Malaysia. (Photograph: Iain R. Taylor.)

From a sample of 978 church nest sites examined in Burgundy, 81.3% were in the bell tower and 18.7% in the roof space (Baudvin 1986). Within the towers and roofs and among the ancient great wooden beams and vaults, these old churches offer a multitude of secure, albeit noisy, nesting places. Most often the birds choose to lay near the highest points, on top of the thick walls of the bell room, presumably to gain

Fig. 10.5. In Holland, some barn owls still nest in windmills. (Photograph: Johan de Jong.)

maximum security against predation by beech martens, *Martes foina*. The birds usually come and go through the wooden slats surrounding the bells. Nesting in churches is uncommon in Britain and north America. The British seem to like tidy churches and most have long since been wired up to exclude jackdaws and pigeons.

In my Scottish study areas, 85.2% of nests were in abandoned or

Fig. 10.6. Abandoned cottages were the most frequently used nest sites in south Scotland. The birds usually nested in the attic space but sometimes used chimneys that had become blocked up with jackdaw nests. (Photograph: Iain R. Taylor.)

seldom used cottages and farm buildings (Fig. 10.6, Taylor 1991b). Within these, the birds most often used attic or loft spaces, usually selecting the darkest corners under eaves or in recesses, or in spaces between the ceiling and attic floorboards. Often they nested down chimneys, frequently evicting jackdaws and using their wool-lined stick nests. The owls normally had to share these buildings with large numbers of jackdaws and sometimes they nested only 2 or 3 m apart, each being able to watch the other during incubation and sharing the same entrances. In the course of a 14-year study, not one of these owls lost a single egg, so presumably the relationship was harmonious. Old dovecots, originally built to house ornamental pigeons or sometimes to provide a winter meat supply, were another frequently used site and again the owls sometimes shared these places with their more normal inhabitants.

Barn owls in southern England frequently nest in haysheds, seeking out places where bales have been loosely stacked so that a deep tunnel has been formed between them (Fig. 10.7 and Fig. 7.4, p. 104), or laying on top of loose hay (Bunn et al. 1982). By contrast, this site is now used infrequently in Scotland: in my areas only 6 out of 501 nesting attempts recorded were among bales in haysheds. The difference is probably related to climate and farming styles. With shorter growing seasons and later springs, much less hay was normally left in Scottish barns by the time the owls started to nest. Bales were usually stacked very tight to keep out rain so the vast majority of haysheds were unsuitable anyway. Also, many were used by shepherds for lambing and fostering at exactly the time in spring when the owls would normally be

Fig. 10.7. Barn owls in Britain sometimes nest between bales in haysheds. This can create problems if the owls lay early and the farmer still needs to remove hay. The difficulty can usually be overcome by providing nest boxes in late winter in sheds that are seen to be frequented regularly by the owls. They usually prefer the box to the bales. (Photograph: Robert T. Smith.)

settling in. To test the significance of this disturbance, in 1981 I placed 10 nest boxes in haysheds that were used in this way by shepherds and 11 in sheds that were not used. In the following 8 years, barn owls bred in nine of the boxes in undisturbed sheds but in none of those in the disturbed sheds. In the ninth year however, two of these were used for breeding, in both cases following a change in work-pattern of the farms which left the sheds unused in the spring.

10.2 Nest site preference

Certain criteria normally have to be met before a site will be accepted by the owls for nesting. Where potential mammalian nest predators such as martens, mongoose or racoons are present more inaccessible locations seem to be selected. Height seems to be important in this: in New Jersey, barn owls will seldom accept artificial nest boxes placed at heights of less than 6 m (Colvin 1984). In the absence of predators, as is the case in Britain, sites almost at ground level may be used and nest boxes at only 2 m are sometimes accepted. Dryness is important and, when there is a choice, the birds seem to prefer dark sites, again possibly to reduce detection by predators, although dark sites must often offer a better insulated microclimate which could also be important. Cavity size is significant and the floor area generally must exceed 0.3 m by 0.3 m and usually is considerably larger.

When nesting in association with humans, relative freedom from frequent and prolonged direct disturbance that flushes the bird out of the nest area is clearly important but extremely difficult to quantify in any precise way. There is also evidence that its significance is not uniform across all barn owl subspecies. North American birds seem to be extremely sensitive especially during the early stages of breeding and until the chicks are no longer brooded by the female (Bruce Colvin, personal communication). In Scotland, I have visited hundreds of nests and trapped adults at the nest at all stages from courtship through to laying, incubation and hatching without causing desertions. My visits were generally at about 10- to 14-day intervals so obviously the birds could tolerate this level of disturbance (Taylor 1991a). In practice, the great majority of sites used in my areas were hardly ever, or never, visited by humans apart from myself. When the owls were in a concealed cavity where they could sit without seeing humans, such as in a roof space, chimney or water tank, they seemed able to tolerate considerable noise and human activity at very close range (Fig. 10.8). On several farms they even showed evidence of habituation to humans and some birds would remain on their roosting perches in barns while people worked only 10–15 m away.

Potential nest sites presumably vary in the extent to which they fulfil these criteria and one might predict that the birds would select sites according to their quality, but the question of nest site preference has rarely been investigated. Bruce Colvin (personal communication) found that when he installed nest boxes high up on the inside of the gable end of barns with access to them only through a hole in the barn wall, barn owls nesting in nearby tree holes almost always quickly transferred into the boxes. The boxes were totally inaccessible to racoons whereas the tree holes were not. Shawyer (1987) reported that in Britain the proportion of barn owls nesting in trees, in comparison with buildings, was highest in the east and declined towards the west. From this he concluded that the owls preferred to nest in buildings in the west as these offered greater protection from rain which also increased westwards. I was able to test Shawyer's conclusion by comparing the use of buildings and trees in relation to their availability within my study areas. No selection was shown; the birds simply used the two types of nest site

Fig. 10.8. Barn owls nested in the roof cornice of this New Jersey farm house (see arrow). Such sites are also used in Europe. (Photograph: Iain R. Taylor.)

according to their availability (Taylor 1991b, Table 10.1). The reason for the low percentage using tree holes was simply the low density of trees with suitable cavities available to the owls. This shortage of tree sites in western Britain has been noted by others (e.g. Seel, Thomson and Turner 1983). Shawyer also suggested that the birds became imprinted on the nest site type in which they were raised. I have recorded six cases where birds reared in tree holes subsequently nested in buildings, or *vice versa*, and two cases where individuals that had already nested in tree holes switched to buildings following the collapse of the trees in winter gales. Colvin's experiences, described above, of individuals moving from tree sites to nest boxes in buildings also suggest that early experience is not involved in site preference. A conclusive testing of this would require rigorous experimentation and is probably only possible with captive birds. Interestingly, similar tests done with American kestrels failed to demonstrate any influence of early learning (Shutt and Bird 1985).

Table 10.1. *Barn owls in south Scotland show no preference for nest sites in buildings compared with those in tree cavities*

	Study areas	Breeding sites		
		No. available	No. used	% used
Buildings	1, 2, 3	123	80	65.0
Tree holes		18	10	55.6
Buildings	4	14	4	28.6
Tree holes		32	11	34.4

Study areas 1–4 as shown in Fig. 1.2, p. 4.

10.3 Summary

In their natural state, barn owls are cavity nesters, depending upon holes in trees and cliff faces but also making use of the nests of other birds such as the hammerkop in Africa. European and North American barn owls particularly, have become highly dependent upon man-made structures, especially agricultural buildings such as barns and storage silos and church bell towers. Nesting in tree cavities remains widespread in parts of the USA and England but is relatively uncommon in continental Europe and Scotland. Barn owls select very large cavities usually in the main bole of the tree and, in America, these tend to be at greater heights than in Europe, probably in response to the predation risk from racoons. Freedom from direct, frequent and prolonged human disturbance that flushes the birds away from the nest area is an important characteristic of suitable breeding sites. North American barn owls are much more sensitive in this respect than European birds.

11 *Courtship and eggs*

S ome species of bird show very little variation in their clutch size, laying much the same number of eggs regardless of habitat or year. In contrast, barn owls are characterised by great variability and clutches ranging from 2 to 18 eggs have been recorded. Variation exists within populations in any given year and individual females can also produce markedly different clutch sizes in different years.

David Lack (1968) suggested that clutch sizes, along with other reproductive features, have evolved so that birds produce, on average, the largest number of surviving young. Yet, many females in some popu-

lations produce clutches smaller than the most productive (Perrins and
Moss 1975). This suggests that for these birds laying more eggs would
entail a penalty of some kind. Female barn owls, in common with many
species, accumulate reserves before laying which are used up gradually
during incubation (Fig. 11.1). Producing more eggs might eat into these
reserves such that the female is less likely to complete incubation successfully, and rearing a large brood of young might also leave the birds
so physically exhausted that their chances of producing a second clutch
or of surviving to breed again in subsequent years would be reduced.
Added to this, young from a large brood might fledge at lower body
weights and have a reduced chance of survival. Thus, it is possible that
a bird such as the barn owl that potentially can lay more than one clutch
in a single season and that has a chance of breeding in more than one
season could rear more young during its lifetime by laying fewer eggs
than the maximum possible at any single breeding attempt.

Drent and Daan (1980) proposed that there is a close interplay
between individual females and their environment such that clutch size
is continuously adjusted to local conditions of food supply. In effect,
they proposed that females behaved in a 'prudent' way and are able to
minimise the risk of damaging their own condition and survival by modifying their clutch size according to their energy or nutrient balance at
laying time. In the last chapter it was shown that barn owl laying dates are
closely correlated with local conditions of food supply. In this chapter, I
will examine relationships between clutch size and food supply but
before doing so will discuss some interesting aspects of courtship, the
eggs and incubation periods.

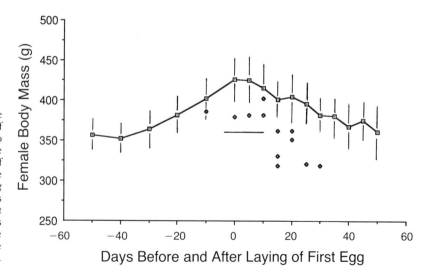

Fig. 11.1. Changes in the mean body weights (± SD) of breeding females in relation to egg laying. The horizontal line shows the average weight of non-laying females during the laying period of breeding females. Individual points show weights of females that abandoned their clutches before hatching: most were substantially below average weights.

11.1 Courtship

Barn owls usually conduct their lives quietly and unobtrusively, and it is difficult for the average observer to gain much insight into their behaviour. However, with the coming of spring and the start of the breeding season they become more conspicuous and whilst their advertising and courtship behaviour could hardly be described as spectacular it has a subtle beauty and more than a hint of mystery.

The first signs of breeding activity can be detected some 6 to 8 weeks before the laying of the first eggs but just as the timing of laying varies greatly from year to year so also does the onset of breeding behaviour. In some years with very high vole populations, the first indications in the Scottish area were evident in early February, but more normally activity begins in the first half of March.

The owls' activity is centred very much in and around the nest site but the precise sequence of events and behaviour varies greatly among different individuals, particularly the males. Some existing pairs may have roosted together at the nest site over winter, but most will have roosted apart and, in the case of the males, usually away from the nest site. I spent many winter nights observing at a number of nests and more often than not the male would visit the site early on in the evening, presumably to maintain his ownership. Usually the first sign of increased activity involves these males spending more time at the nest site at night and a gradually increasing tendency to call using their long drawn out screech. This is a highly variable call, sometimes powerful, long and clear when it has a distinct vibrato or tremulous effect, at other times weak and almost hoarse. Presumably it has two functions: to advertise ownership of a site to other males and to attract mates. Unmated males use this call more persistently and more loudly than mated males. Among the latter, its frequency increases immediately before laying of both first and second clutches (Buhler and Epple 1980). Presumably this is to exert the maximum deterrent to other males and so minimise the chances that an intruder will mate with the female during the most vulnerable period of fertility and egg formation. Usually the call is delivered from the building or tree containing the nest chamber but they also employ a ritualised advertising flight display in which the screech call is also given. In this, the males make brief sorties over the fields immediately around the nest using somewhat stiffened, deliberate wing beats.

As the season advances, the males spend more and more time around the nest and in the company of their females. Each time I visited nest sites during the day I recorded whether the female or male or both were

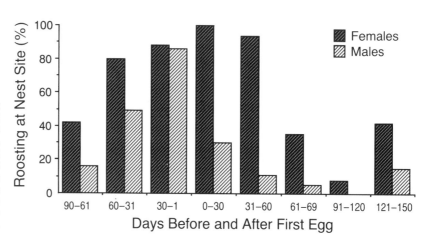

Fig. 11.2. Seasonal changes in the percentage of male and female barn owls roosting at the nest site, south Scotland. Immediately before and during laying most males roost alongside their females. This is probably mate-guarding behaviour. Data from 3743 visits to 740 nests over 14 years.

roosting there. The percentage of males that were sitting beside their females increased to reach a peak in the 2 weeks before laying and during laying, when almost all males were at the nest. Within a few days after clutch completion, most had moved out, returning to their more usual roost sites (Fig. 11.2). This quite distinct behaviour probably functions as mate guarding, to try to prevent other males from fertilising the female. Similar behaviour occurs in a wide range of birds (Birkhead and Muller 1990) but tends to be less pronounced among birds of prey. Active hunters that must search widely and on their own to catch difficult prey are less able to accompany and guard their mates at all times than are other species. By spending all of the daylight hours with their mates, male barn owls are at least able to eliminate the possibility of another male sneaking in 'in broad daylight', which they otherwise could easily do. However, they also increased the amount of time spent at night with their mates. Four radio-tagged males studied in detail spent only 26% of the night-time away from the nest in the period immediately before and during laying (8 nights and 27 hours observation), whereas in the 10 days following clutch completion they were away from the nest for 72% of the night-time (8 nights and 32 hours observation). This of course still leaves some time for other males to come in and mate with the female. It is perhaps not surprising then that male barn owls copulate very frequently with their mates (Buhler and Epple 1980, Bunn *et al.* 1982, Taylor, personal observation). Copulation occurs before he starts on his first foraging trip of the night and is repeated many times when he returns with prey or apparently completely spontaneously. Given that he must spend time hunting, he really does all he can to reduce the chances of bringing up some other male's offspring. Nevertheless, it would be intriguing to do some DNA fingerprinting to find out just how successful his efforts are.

Food presentation is an important part of courtship and probably also in the initial attraction of females. Unmated males usually have a stock-pile of dead prey at their intended nest sites; in one case I recorded no less than 136 dead, and mostly decomposing, items. Fresh prey are usually added each night and, although I have never been in the right place at the right time to see it, I assume that when a female is attracted in by his screeching she is very quickly presented with one of these. Perhaps she is able to judge the quality of the male or of his home range by the number of items he has on show. Giving and receiving food probably strengthens and maintains the pair bond. However, more importantly it is responsible for the rapidly increasing body weight of the female before laying, providing reserves for egg production and for use during incubation (Fig. 11.1). For about 10 days or so before laying she remains at the nest and her mate provides all her food supply. Deliveries reach an extremely high level during laying (Fig. 12.5, p. 172), so much so that they sometimes greatly exceed the female's requirements and a surplus builds up beside her. In the Esk study area, clutch size was related to the body weight of the female immediately before laying (Fig. 11.3). This could have been because heavier females had more reserves for laying or because they received a higher day-to-day level of food supply from their mates that could then be channelled into egg production. Either way, this stresses the importance of the male's role as a food provider.

Before laying, there is a great deal of ritual to be gone through, the significance of which is not yet fully understood. Sexual chases occur frequently, with male and female flying in and around the nest site at high speed and swerving, twisting and turning through open doors and windows or over the rooftops, all the time screeching loudly. Bunn *et al.* (1982) have described a more sedate display in which the male follows

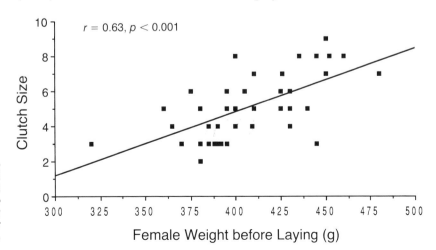

Fig. 11.3. Relationship between clutch size and female body weights measured from 5 days before laying to the start of laying. Esk study area, south Scotland. Heavier females laid larger clutches.

the female, flying about 2 to 6 m above and slightly behind. The male has a ritualised display in which he repeatedly flies in and out of the nest building using stiffened, shallow, rapid wing beats and calling repeatedly. I have seen this only when the female is present and its function may be to encourage her to enter the nest site.

The birds have an extensive vocal repertoire. Apart from the screech call, the female has a snoring-like vocalisation, used from early courtship onwards, described in detail by Buhler and Epple (1980) and Bunn et al. (1982). A very similar sound is highly characteristic of nestling barn owls and is associated with hunger and feeding and probably stimulates the parents to provide food. The same may be the case with the female, encouraging the male to feed her. There are also special calls used by males and females during copulation. There is a great deal of contact during courtship: bill fencing, mutual preening and 'cheek-rubbing', usually accompanied by tongue clicking, various squeaks and twittering and a snoring hiss.

Occasionally, the male succeeds in attracting more than one female. Many times during the winter months I have discovered two females and sometimes even three roosting together at nest sites. Usually these are the resident females and first year visitors, who normally move away again before the breeding season. However, even during the breeding season, new females sometimes appear, spending anything from a few days to a week or two roosting alongside the incubating female. Very occasionally this association goes a stage further and the second female is mated by the male and produces a clutch. I have recorded this bigamy in 7 out of 419 nesting attempts in the Esk study area and was first alerted to it by seeing the same colour-marked male carrying food to two nests. Usually the two females were in separate nest sites up to about 1 km apart and one of them received much less food that the other and consequently was less productive. The single notable exception involved two females who laid clutches of four and five eggs only centimetres apart so that during incubation they were nestling into each other. From each clutch only one young was fledged.

11.2 Eggs

Barn owl eggs are oval and when newly laid are beautifully white but in most nests they gradually become discoloured from the broken pellet debris that surrounds them (Fig. 11.4). One exceptional clutch I observed in Scotland had all of five eggs finely flecked with dark brown markings. Infertile eggs may be incubated for as much as 6 weeks and can become very grimy before being abandoned. The average measure-

Fig. 11.4. A clutch of barn owl eggs. Clutch size in Europe can vary from 2 to 12 depending upon prey abundance and habitat type. (Photo: Robert T. Smith.)

ments of Scottish eggs of the *alba* subspecies were 42.1 × 31.8 mm (*n* = 62) and most weighed between 16 and 20 g, with an average of 17.9 g at laying. A dark-breasted *guttata* subspecies has eggs averaging 39.2 × 30.9 mm in size and 17.5 g in weight (Schonfeld and Girbig 1975). Those of the North American subspecies are considerably larger, at 43.1 mm × 33.7 mm and averaging 26.6 g (Marshall, Hager and McKee 1986).

Egg weights of birds in general, are very closely related to female body weights (Rahn, Paganelli and Ar 1975). The relationship is allo-metric, that is to say, egg weights are related logarithmically to female weights so that as body size increases the eggs form an ever-decreasing percentage of female weight. Barn owl egg weight as a percentage of female weight, is 4.9%, 4.9%, and 4.8% in *alba*, *guttata* and *pratincola* respectively. Barn owl eggs immediately give the impression of being remarkably small. To investigate this further, I examined the relationship between female body weights and egg weights of all European owls from sources given in Cramp (1985), in addition to those for barn owls above (Fig. 11.5). In this analysis, female weights were taken as those

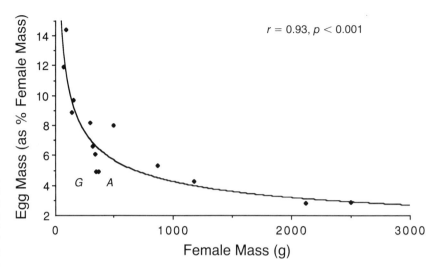

Fig. 11.5. Egg weights (as % of female weight) of European owl species, in relation to female weight. Barn owls produce markedly smaller eggs compared with other species. G is *T. a. guttata*; A is *T. a. alba*.

before the breeding season increase. For the 13 species, the relationship was described by the following equation:

Log egg weight = Log 0.726 + 0.593 Log female weight ($r = 0.93$)

Were barn owls to lay eggs equivalent in size to those of other species, they would weigh (for *alba*) 24.6 g instead of the 17.9 g actually produced.

If food supply in the form of energy or specific nutrients is limiting at the time of laying, birds could respond by adjusting clutch size or by modifying egg quality or size. In an evolutionary sense, the options will depend upon the importance of other factors: there would be considerable disadvantage in a wading bird producing more but smaller eggs if the resulting small chicks suffered a higher predation rate or were less able to move through vegetation. However, if there is no such disadvantage, natural selection might favour the production of larger clutches of smaller eggs, which inevitably would involve a longer post-hatching development period. Thus, cavity nesting species which are more secure from predation tend to lay more but smaller eggs than open nesters (Lack 1966). When only those owl species that, like the barn owl, are quite clearly cavity nesters were included in the analysis, the relationship between egg weight and female body weight was described by the slightly different equation:

Log egg weight = Log 0.61 + 0.631 Log female weight ($r = 0.96$)

This predicts that the barn owl (*alba*) should lay eggs weighing 25.5 g. Thus, the exceptionally small size of barn owl eggs cannot be explained simply by the hole nesting habit. This leaves two other possibilities. Egg production may be more limited by the day-to-day food supply received by the female during laying than in other owl species so that egg weight is the maximum possible. In view of the surpluses of fresh prey usually found in nests at this time, this explanation seems unlikely. Alternatively, small egg size might be an evolutionary strategy whereby using the same body reserves, the birds are able to lay more eggs and ultimately rear more young. If barn owl eggs were of the predicted weight, a clutch of five eggs would weigh 125 g. This would equate to a clutch of about seven eggs at the weights actually produced by barn owls. The difference is therefore considerable. Laying more eggs would only be possible if the birds could expect high prey availability when rearing young or if they have evolved ways of coping with the feeding of larger broods of young.

11.3 Incubation

Barn owls usually lay their eggs at 2- to 3-day intervals although longer gaps are sometimes encountered. The most extreme case I came across involved a first year female who was at the lower end of the usual weight range a few days before laying. I next visited the nest almost 7 weeks later to find two young, aged about 12 to 14 days, and two eggs which were chipping and hatched 1 to 3 days later. It seems unlikely that their incubation period could have been prolonged by so much, so probably there was about a 2-week interval between laying the first and second pairs of eggs. I recorded several other instances where there were intervals of 3 to 7 days between successive eggs. Most of these were associated with spells of unusually adverse weather conditions: extremely low temperatures, snowfall, sleet or prolonged rainfall during laying. It is possible that the amount of food the males were able to deliver to their mates was reduced at these times and, if so, this suggests that the day-to-day intake of food was important for egg formation.

Some days before laying the females lose their feathers in the belly region and develop a brood patch, an area of bare skin which becomes richly supplied with blood vessels. Of many hundreds of male barn owls I trapped during the incubation period none had developed a brood patch. Males are therefore unable to incubate properly and this task is performed entirely by the female. There are however a few records of males apparently sitting on eggs for brief periods (Baudvin 1975). Incubation starts immediately after laying of the first egg, which results in the eggs hatching a 2- to 3-day intervals. Incubating females tend to sit very tightly and some older birds that had become well habituated to my visits would even remain and allow me to nudge them aside to check their clutch sizes. Periods spent off the eggs are short and infrequent, usually less than four times each day.

There is some variation in estimates of incubation periods but, for the *alba* and *guttata* subspecies, most eggs hatch after 30 to 32 days with a range from about 29 to 34 days (Muller 1989, Bunn *et al.* 1982 Taylor, personal observation). Compared with other owls this is very prolonged. In Europe, only the eagle owl (*Bubo bubo*) has a longer incubation; the barn owl's incubation is about the same as that of the snowy owl (*Nyctea scandiaca*) whose eggs are about three times heavier. There is usually a close relationship between female body weight and incubation period. For several hundred species sampled, this is given by the equation:

Log incubation period = Log 26.76 + 0.239 Log female weight

This predicts an incubation period for barn owls of only 23 days (Rahn and Ar 1974). Owls in general and especially hole nesting species have relatively long incubation times, but, compared with other European cavity nesting owls, the barn owl still has an incubation lasting some 3 to 4 days or about 10% longer than expected on the basis of female weight.

11.4 Incidence of non-laying

Under extreme conditions of food shortage, female owls may sometimes respond by not laying at all. Among European species, this has been recorded in Ural owls, *Strix uralensis* (Saurola 1989), and tawny owls, *Strix aluco* (Southern 1970). In the Scottish barn owls, definite non-laying, when both male and female were caught and known to be alive during the breeding season, was not common and occurred almost exclusively during the decline and low phases of the vole cycle (Table 11.1). The weights of these non-laying females were markedly lower than those of breeding females during the main laying period, suggesting that the cause of non-breeding was a failure to accumulate adequate body reserves (Fig. 11.1).

In 1991, an interesting case was observed of a pair that had bred successfully together for the previous three seasons. When trapped in April, the male was clearly recovering from a fractured left tarsus. Prey abundance was high in early spring and all other pairs in the area laid large clutches. However, the female of this pair failed to gain weight above her normal early spring level and did not lay. Presumably the male's ability to capture prey for the female was impaired by his injury.

During low vole years, many nests in the study area were occupied

Table 11.1. *The incidence of non-laying among barn owl pairs in the Esk and Dee study areas between 1980 and 1992*

Year	Esk study area		Dee study area	
	No. of breeding pairs	No. of non-breeding pairs	No. of breeding pairs	No. of non-breeding pairs
1980	39	0	16	0
1981	55	1	21	0
1982	42	6	17	1
1983	23	1	10	1
1984	34	0	16	0
1985	21	1	10	1
1986	13	1	8	0
1987	22	1	–	–
1988	20	2	–	–
1989	16	0	–	–
1990	32	0	–	–
1991	28	2	–	–
1992	24	0	–	–

by single, non-breeding individuals, both male and female. This could be established only by having a detailed knowledge of ringed birds. Several other studies (e.g. Muller 1989) have also established with certainty that some pairs do not breed in years of low food supply, but, because of the difficulty of distinguishing between single non-breeders and pairs, the relative importance of each could not be established.

11.5 Seasonal trends in clutch size

In the Scottish population there was a strong correlation between annual average clutch sizes and annual average laying dates: years with early laying were also years of high clutch sizes (Fig. 11.6a). Within individual years there was also a close relationship in that early laying pairs tended to produce the largest clutches and as the season progressed clutch size declined (Fig. 11.6b). However, markedly different results have been obtained for some populations in continental Europe and Africa. In Germany, the number of eggs in first clutches was much higher in June and July than in April and May when most pairs laid. Second clutches that were laid mostly in July and August were significantly larger than first clutches laid on average 2.5 months earlier (Schonfeld and Girbig 1975). Similar results have been obtained from studies in France (Baud-

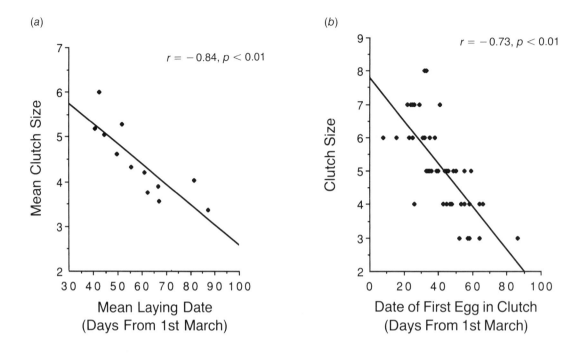

Fig. 11.6. Seasonal declines in clutch sizes of barn owls in the Esk study area, south Scotland. (a) Annual mean values for the whole population from 1980–1991. (b) Values for individual females, 1981.

vin 1986). In Mali, clutch sizes were smaller at the start of the breeding season, reached a peak in mid-season and declined towards the end (Wilson *et al*. 1986).

Seasonal declines in clutch size have been recorded in many birds, including diurnal raptors such as European kestrels and sparrowhawks (Village 1990, Newton 1986), and seem at first consideration to present something of an enigma as weather conditions and prey abundance commonly improve as the season progresses. Suggested explanations for seasonal declines are that later laying birds are poorer foragers or in less good habitat so that their food intake is limiting (e.g. Newton and Marquiss 1984) or that natural selection has favoured individuals that lay fewer eggs later in the season even though food supply may be high enough to produce more eggs (Drent and Daan 1980). The latter implies that the laying of larger clutches later in the season would result in fewer young being fledged than from smaller clutches or would result in reduced survival of the adults. It also requires some external control of clutch size, perhaps by day length changes. Current evidence does not indicate a clear preference for either suggestion. For example, experiments with European kestrels showed that later laying pairs did not produce larger clutches when provided with extra food before laying, supporting the second hypothesis (Meijer 1980). However, individual kestrels that were forced to produce repeat clutches laid more eggs than in their first attempts and more than natural clutches at that part of the season (Village 1990), supporting the first idea.

Do the different seasonal trends found in the three barn owl subspecies in Scotland, central Europe and Africa really indicate different types of response or are the owls possibly reacting to food supply in a similar way but in areas with different seasonal patterns of availability? Vegetation growth probably tended to reduce the availability of field voles to the Scottish owls as the season progressed and the intense territorial activity of the voles which exposed them to predation almost certainly peaked early in the spring. The annual increase in vole numbers through breeding did not start until about mid-June. It is thus possible that prey availability declined towards mid-summer and the seasonal decline in clutch size was correlated with a declining food supply. For the African barn owls, vegetation cover would have declined steadily during the dry season as laying progressed, but so also would have the absolute numbers of their rodent prey whose main breeding periods were in the latter part of the rainy season. Thus, their food supply was probably low at the start of laying when the vegetation was still high, peaked some way through and declined towards the end as prey numbers fell. Clutch sizes again could have been correlated directly with food supply. Continental European barn owls prey mainly upon the common vole which, because of its burrowing habit, is not restricted to long vegetation. There might thus be no strong seasonal decline in their availability to the owls as a result of vegetation growth. The owl's food supply might then increase seasonally with increasing vole populations and once again clutch size may simply have been related to availability. It is possible that barn owls differ fundamentally from species such as the kestrel. Probably, barn owls evolved in savannah or lightly wooded habitats in generally hot environments where vegetation growth and the breeding seasons of small mammals were somewhat unpredictable in response to variable rainfall patterns and where the opportunities for the owls to breed were not clearly predicted by day length changes. Barn owls might therefore respond directly to changes in prey availability rather than to changes in day length; they might simply not be photoreceptive, even when living in temperate climates. There is clearly a good case for some controlled laboratory trials to elucidate this problem.

Why should many pairs in Europe and Africa choose to lay small clutches early in the season when by postponing breeding by a month or two, they could presumably produce larger clutches? Perhaps the main reason is that early laying pairs are often able to produce a second clutch, so that they eventually fledge more young and young fledged early in the season probably survive better than those fledged later. Natural selection would thus favour early laying even if it means laying fewer eggs in the first clutch.

11.6 Second clutches

In the Scottish population, second clutches, confirmed by trapping ringed females, occurred almost exclusively in peak vole years but even then involved only a small number of pairs and never more than 15% of the population. Pairs that produced second clutches invariably laid their first clutches early in the season, in March or early April, and the average interval between starting the first and second clutches was 98 days ($n = 21$). Thus the new clutches were begun about a week or two before the first brood was fully fledged. Second clutch sizes were smaller than first clutch sizes of the same females (means of 4.6 compared with 6.3).

In continental Europe, second clutches are most frequent in the central area of eastern France, Germany and Switzerland and less frequent in western and Mediterranean areas (Muller 1991). The region of highest incidence is also the zone where the owls depend most upon common voles. Yearly variations in vole abundance were not measured directly in the studies conducted in these areas but nevertheless peak and low abundance years could be identified. In Germany, for example, the laying of first clutches was earliest and the incidence of second clutches greatest in peak vole years when up to 64% of pairs laid twice. Averaged over 7 years, 32% of pairs produced two clutches and three females were recorded laying three clutches (Schonfeld and Girbig 1975). Very similar results have been obtained from barn owl populations in France (Baudvin 1986, Chanson et al. 1988, Muller 1991).

The average number of eggs in second clutches was almost always substantially greater than in first clutches. For example in the peak vole year of 1971, Schonfeld and Girbig (1975) found average clutch sizes of 5.48 and 8.23 eggs for first and second clutches respectively. Unfortunately, only averages for the whole population were given with no subdivision of first clutches for those produced by single- and double-brooded females. However when this separation was made by Muller (1991), the much larger size of second clutches was still apparent.

The pattern of second clutches in Mali was similar to that for Europe, with a greater incidence in years of high prey abundance, following the early laying of first clutches. Second clutches were again larger than first clutches (Wilson et al. 1986). In Malaysia, most females seem to lay at least two clutches in most years but there seems to be no significant difference between them in the number of eggs laid (Lenton 1984a,

John Duckett, personal communication). Some North American barn owls also produce second clutches in high prey years (Bruce Colvin, personal communication) but details of their incidence and clutch size have not yet been published. In the southern, drier states such as Texas, second clutches have not been recorded (Otteni *et al.* 1972), a situation similar to that found in Mediterranean Europe.

11.7 Clutch size and annual variation in prey abundance

The Scottish owls experienced annual, cyclic variations in their food supply in spring and average annual clutch sizes for the whole population were closely correlated with these changes, following an exactly parallel cyclic pattern (Figs. 11.7, 11.8, 11.9). At vole peaks, first clutches aver-

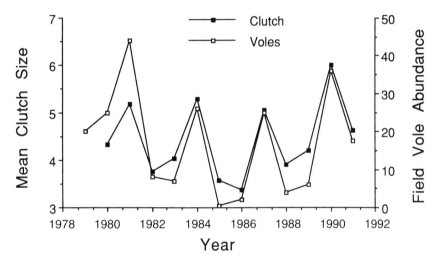

Fig. 11.7. Annual average clutch sizes show a cyclic pattern clearly in synchrony with the field vole cycle. Esk study area, south Scotland.

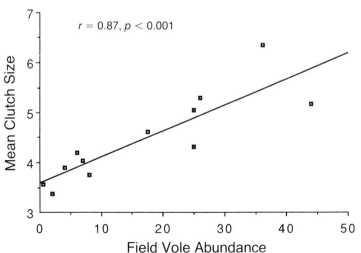

Fig. 11.8. Average clutch sizes in the Esk study area were closely correlated with spring field vole abundance.

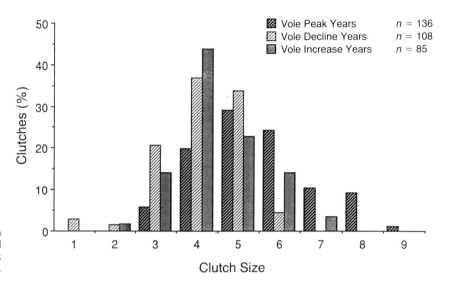

Fig. 11.9. The range of clutch sizes in peak, decline and increase vole years in the Esk study area, south Scotland.

aged between five and six eggs whereas in decline years this fell to between three and four eggs. When second clutches are added to this, overall egg production was increased further by about 0.3 to 0.5 eggs per pair in peak years. The largest clutches recorded were nine in peak years and six in decline years (Fig. 11.9).

In Germany, where cyclic populations of common voles were the main food, first clutches averaged 5.5 eggs in a peak vole year but only 3.2 and 4.6 in two low vole years (Schonfeld and Girbig 1975). Most other European studies have demonstrated annual fluctuations in clutch size of similar magnitude (Kaus 1977, Baudvin 1986, Chanson *et al.* 1988, Muller 1991). African barn owls in Mali laid an average of 6.2 eggs in a year of high rodent numbers and 5.5 in a year of low numbers (Wilson *et al.* 1986).

Less yearly variation has been observed in North American populations. In Texas, average clutch size remained at around five eggs for 5 consecutive years and then fell to 3.8 for 2 years following a widespread decline in rodent numbers (Otteni *et al.* 1972). This owl population had a highly diverse diet and showed considerable year-to-year switching of prey species. This may have been responsible for the greater stability of egg production. A population in Utah produced average clutches close to seven eggs in 5 out of 6 years (Marti and Wagner 1985). In this area, vole numbers did not vary cyclically (Carl Marti, personal communication).

There is therefore a great deal of evidence from many parts of the world that barn owl clutch sizes are closely correlated with food supply.

Table 11.2. *Average clutch sizes in conifer plantation habitats, sheepwalk and low pastoral farmland, Esk study area, south Scotland*

Year	Vole population[a]	Average clutch size		
		Plantation	Sheepwalk	Low farmland
1980	I	4.7	5.1	3.4
1981	P	6.2	5.7	4.1
1982	D	4.2	3.5	3.7
1983	I	4.9	4.1	3.6
1984	P	6.6	6.0	4.2
1985	D	3.7	3.1	3.7
1986	I	4.5	3.3	3.0
1987	P	6.2	5.2	4.3
1988	D	4.0	4.2	3.7
1989	I	4.5	4.2	4.1
1990	P	6.7	5.3	6.0
1991	D	5.0	4.4	4.6

P, peak vole year; D, declining vole numbers; I, increasing vole numbers.
[a]Vole population changes.

11.8 Clutch size in relation to habitat and female age

Annual average clutch sizes differed significantly among the three habitat types occupied by the owls in the Scottish area: pairs nesting in plantations produced the largest clutches and those in lowland, pastoral farmland produced the lowest (Table 11.2). Thus, egg production was positively related to the density of suitable foraging habitat around the nests. However, there was an interesting subtlety in these differences over the course of the field vole cycle. Differences between habitats were greatest in vole peak years when pairs nesting in plantations produced averages of six to seven eggs and those in low farmlands only around four eggs, with the exception of 1991 when the latter group laid particularly large clutches of about six eggs. By contrast, in vole decline years, there was very little difference between these habitats. In effect, pairs nesting in plantations showed wide, cyclic fluctuations in their egg production, whereas those in low farmland, although also quite distinctly cyclic, showed much less pronounced fluctuations. There may be two reasons for these differences. Firstly, there was some evidence that the amplitude of vole cycles was much greater in plantations than in the farmland. Next, the prey spectrum of the farmland nesting pairs was wider than the plantation nesting pairs so that as vole and shrew numbers declined they could switch their predation slightly to wood mice, an option which was not available to those in the plantations. These results are somewhat similar to the situation described above, in Texas, where a wider prey spectrum and prey switching was associated with greater stability of egg production.

First year breeders made up a substantial proportion of the Scottish breeding population in peak vole years but at other times there were too few of them to allow statistical comparisons of clutch sizes in relation to female age. However, at vole peaks, and controlling for different habitats, first year females did not produce significantly fewer eggs than older females.

11.9 Summary

During courtship and until clutch completion, male barn owls form a close association with their mates, roosting with them during the day and minimising their night-time absences. This behaviour may be mate guarding, which, along with very frequent copulation, may have evolved to reduce the chances of other males fertilising their females. Food presentation is an important component of barn owl courtship which serves to increase the female's body reserves before laying.

Compared with other cavity nesting owls, barn owls lay eggs that are about 30% less in weight. Low egg weight may be an adaptation to allow the laying of larger clutches and hence to increase production. Incubation periods are also longer than predicted from female body weight. Clutch size declines seasonally in Scotland, but not in continental Europe. Second clutches which are usually larger than first clutches are also more frequent in the continental areas. These differences may be caused by a dependence upon prey species which differ in their seasonal patterns of availability. There is extensive evidence that barn owl clutch size varies in relation to food supplies at the time of laying, either in response to yearly variations in prey abundance or to differences in habitat quality.

12 *Production of young*

The ultimate goal of all animals is to produce the maximum number of offspring during their lifetime that in turn survive to become members of subsequent generations. Productivity can therefore be considered at the level of individual animals, to understand how it varies among individuals and to elucidate the environmental factors that are responsible for variations. Alternatively, it can be examined at the population level, as one of the main components in population dynamics. From the viewpoint of understanding predator–prey relationships, both levels are of great interest but for conservation management the population level is the more important.

If estimates of productivity are to be used in the study of population dynamics, it is important that they are representative of the population in question, and there are many sources of possible error. For example, studies might over-emphasise the contribution of pairs that breed in unusually secure places, such as church towers, or they may miss those that failed very early in their breeding attempt or those that breed late. Results from exceptionally good habitats or special conservation areas may well not be applicable to wider areas. Also, it may seem an easy matter simply to visit nests and count the number of young fledged.

However, young barn owls are great wanderers and can secrete themselves into the smallest cracks and crevices within buildings and tree cavities and so easily escape detection, leading to an under-estimation of their number.

To reduce the possibility of errors from such sources, in my studies I selected areas that were natural ecological units, the entire catchments of river systems, so that all habitats and altitudes used by the owls were included, and I attempted to locate all breeding pairs each year. During the winter months, I examined the buildings in which the owls nested in great detail so that I was aware of all the places in which chicks might hide. I tried to visit the nests within a few days of fledging and then again about 2 or 3 weeks later, so that I could be sure of the number of young raised.

12.1 Hatching

A few days before hatching the young can clearly be heard calling within the egg and it seems that most break clear of the shell overnight (Bunn *et al.* 1982). Their efforts to chip through using the specially developed egg tooth may be assisted by their mothers who sometimes break off pieces of shell or remove fragments of membrane (Buhler 1970). Broken shells are normally not removed but become pushed to the edge of the nest area and unhatched eggs are also usually left and can persist for many weeks.

Overall hatching success does not seem to vary greatly among different populations: 71% in Germany (Schonfeld and Girbig 1975), 82% in France (Baudvin 1986), 83% in Mali (Wilson *et al.* 1986) and 69% in Malaysia (Lenton 1984a). Precise details of hatching success were obtained for 8 years in the Scottish study area. There was a significant effect of habitat: 76% of the eggs of low-farmland nesting pairs hatched compared with 85% of the eggs of those that nested in plantations and sheepwalk. Pesticide use was negligible in the low farmland as it was almost entirely pastoral, so this seems an unlikely explanation for the difference. It may be that some of the eggs of low farmland birds were of poorer quality, related to differences in food supply. Average hatching success varied considerably from year-to-year in all habitats but these variations were not correlated with either vole abundance or weather conditions during incubation. This contrasts with results for some other species such as the tawny owl in which hatching success is strongly correlated with rainfall during incubation (Southern 1970). Complete hatching failure (excluding seven infertile clutches) occurred in only 16 (3.8%) of 419 clutches examined in the Esk study area between 1979 and 1991. In the great majority of these cases, the incubating females had body weights considerably below the average for the whole population and abandoned their eggs (Fig. 11.5, p. 156), so food shortage was probably the cause. A sample of 38 nests were studied in much greater detail in 1980 and 1981 by marking eggs according to their order of laying. Of the total 182 eggs laid, 40 (22%) failed to hatch and of these failures, 33 (82.5%) were the last or penultimate eggs in their clutches.

12.2 Chick growth and development

Chicks of the *alba* and *guttata* subspecies weigh about 13 g to 14 g at hatching. They have a sparse covering of early down feathers (the neoptile) which does not adequately insulate against heat loss, but after about 12 days or so the main down covering (mesoptile) begins to grow and becomes fully developed by around day 20. Until then the young are unable to regulate their own body temperature and are brooded by the female (Fig. 12.1). She leaves them alone in the nest only after the youngest is about 12–16 days old. Thereafter, they still huddle together closely, presumably to minimise heat loss and to ensure that as much energy as possible is channelled into growth (Fig. 12.2).

At first, the chicks have their eyes tightly closed and their mothers stimulate them to feed by moving small fragments of prey across their bills. They gain in weight rapidly, achieving their highest daily rate of increase of 12–14 g per day (in *alba* and *guttata*; 22 g per day in *pratincola*) between 16 and 19 days old (this study, Ricklefs 1968, de Jong 1991). Maximum weight is reached by about 36–40 days. This is usually in excess of adult weight and a significant decline normally occurs before fledging although in years of reasonable or good food supply many are still above adult weight when they leave the nest (Fig. 12.3). Growth of the skeleton takes place over much the same period as the increase in body weight; the tarsus for example, reaches maximum length by about day 40 (Fig. 12.4).

The flight feathers do not start to develop until quite late on. Around the 10th day, the primary quills begin to appear and the feathers start to erupt from them by about day 20. Thereafter, they grow at a rate of between 4 and 5 mm a day and are fully developed when the chicks are around 70 to 75 days old (Fig. 12.4). Variations in feather growth rates from day-to-day or among different young in the brood are much less pronounced than variations in weight. Wing feather growth thus provides an accurate estimate of a chick's age particularly between 20 and 60 days, whereas body weight is not particularly reliable. Individual

Fig. 12.1. Barn owl chicks less than 2 to 3 weeks old have insufficient down to regulate their own body temperatures and must be brooded by their mothers. (Photograph: Robert T. Smith.)

weighings depend greatly upon the time since the chick's last feed and the quantity of prey delivered and, over a longer time period, average daily weight increments and also weights before fledging can depend upon prey abundance and the number of young in the brood. In African barn owls, the young that hatched later in broods tended to gain weight less rapidly than earlier young, especially when between 20 and 30 days of age, and weight gain was greatest in broods hatched in the middle of the breeding season rather than earlier or later (Wilson *et al.* 1986).

Fig. 12.2. A brood of young barn owls more than 4 weeks old, well insulated by thick down plumages. (Photograph: Robert T. Smith.)

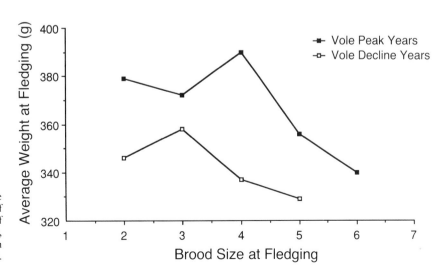

Fig. 12.3. Comparison of the average weights at fledging of barn owl young in years of high and low vole populations, in relation to brood size, south Scotland.

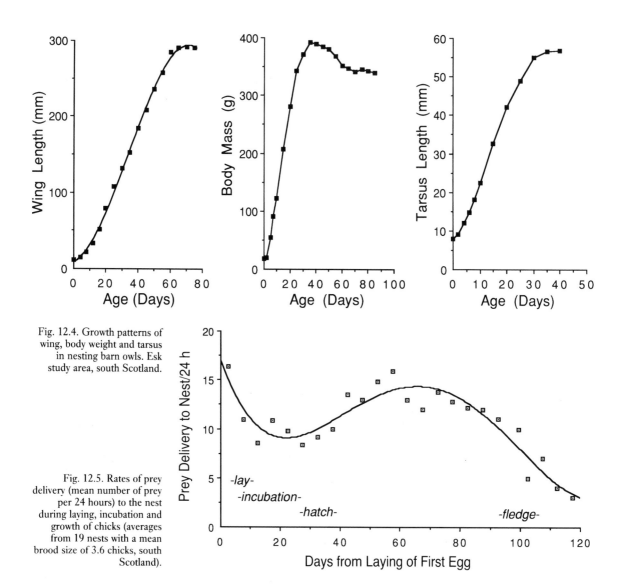

Fig. 12.4. Growth patterns of wing, body weight and tarsus in nesting barn owls. Esk study area, south Scotland.

Fig. 12.5. Rates of prey delivery (mean number of prey per 24 hours) to the nest during laying, incubation and growth of chicks (averages from 19 nests with a mean brood size of 3.6 chicks, south Scotland).

Weights at fledging (60 to 65 days) of the Scottish owls depended on brood size and were significantly greater in peak vole years than in years of low vole abundance (Fig. 12.3).

Rates of prey delivery to the growing chicks were studied at 19 nests in Scotland using automatic recording equipment (Langford and Taylor 1992). This gave a reliable estimate of prey delivery rates but, because young chicks less than about 25 days sometimes did not eat all of the prey brought in, it only gave a good estimate of actual food consumption after the young were about 30 days old. Prey delivery increased from around 8–10 items per 24 hours during incubation when the female was being fed on the nest, to a peak of about 14–16 items when the chicks were about 30 days old (Fig. 12.5). On average, this gave 4.2 items per chick each day. Thereafter, rates declined but some prey were still being delivered when the youngest chicks were about 85 days old. At peak consumption, between 30 and 40 days old, the estimated daily food intake per chick was between 98 and 108 g. Johan de Jong (1991) has estimated that the energetic demand of young Dutch barn owls peaks at about 40 days when their metabolised energy intake was around 450–500 kJ/day. The equivalent figure for the Scottish birds was somewhat higher at about 540 kJ/day but as they were some 12% heavier than the Dutch birds, the two estimates, per gram of body mass, were very similar.

12.3 Mortality of chicks

Barn owl eggs hatch at intervals of about 2 to 3 days, resulting in a considerable age and size hierarchy among the chicks. Such asynchronous hatching is quite widespread in different bird species and its ecological significance has received considerable attention. The earliest explanation (Lack 1954, 1968) proposed that this was an adaptation whereby brood size could be adjusted to prevailing conditions of food supply. Should food supply decline unpredictably, perhaps because of a sudden crash in prey populations or adverse weather conditions, the size hierarchy would ensure that the largest, oldest young would obtain whatever food was provided, whereas the youngest would lose out. Thus, at least some might survive. Were all the chicks to be of the same age and competitive ability, all might perish. This idea was followed by others, for example Hussel (1972) suggested that asynchronous hatching would avoid simultaneous peak food demands of all the chicks in a brood. Spreading the load could enable the parents to rear more young. Clearly, this could be of greater significance in the barn owl which can sometimes rear eight to ten young than in species that produce only two or three offspring.

These and other hypotheses are difficult to separate using evidence from field observations alone and attempts to investigate the problem by artificially manipulating the ages of young in broods have produced very mixed results (see Slagsvold and Lifjeld 1989). Nevertheless, it is interesting to examine the details of barn owl chick mortality in the light of these ideas. Were Lack's suggestion to operate, we should expect that chick mortality would be related to food supply and that the youngest chicks would suffer the highest death rates.

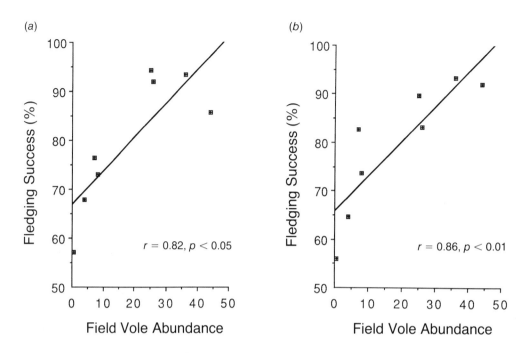

Fig. 12.6. Mortality of barn owl chicks in relation to vole abundance, in (a) young plantation and (b) enclosed farmland habitats, Esk Study area, south Scotland. A similar trend was found among pairs nesting in sheepwalk habitats.

In the Scottish area, chick mortality varied greatly from year-to-year in all habitats and these variations were indeed correlated with prey abundance (Fig. 12.6). In peak vole years, only about 5 to 10% of the young died between hatching and fledging, whereas in the years of lowest vole abundance up to almost 45% of the chicks died. In peak years, the few deaths that did occur involved only the last or second last hatched chicks, whereas all those hatched earlier survived. By contrast, in the low vole years, the mortality of first and last hatched young was 26% and 94% respectively. Baudvin (1978) found that 80% of the chicks that died in his French study were also the last or second last hatched and, in Mali, Wilson et al. (1986) found that the earliest hatched had the lowest mortality rates. Thus there seems widespread agreement that mortality falls preferentially upon the younger chicks and that a higher proportion of them die when food is in short supply. These observations are not in themselves an adequate test of Lack's ideas but they certainly do not contradict it.

The age at which chicks die is also of interest. In the peak vole years in Scotland, almost 70% of all deaths occurred before the young were 10 days old (Fig. 12.7). In most of the nests concerned, there was a substantial surplus of stored prey available when these chicks died and it is hard to accept that all died of food shortage. It is possible that some of these chicks had hatched from poorer quality last eggs in the clutch

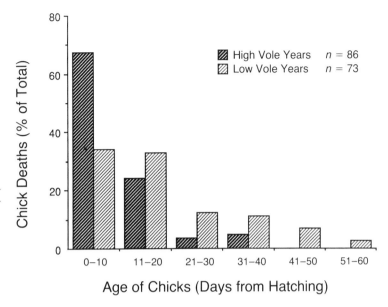

Fig. 12.7. Age of death of barn owl chicks in years of high and low field vole abundance, Esk study area, south Scotland. Most die during the first 20 days or so, but when prey is scarce many continue to die of starvation even to the point of fledging.

and they were simply not strong enough at hatching to stand up and beg for food properly. Last laid eggs in clutches were frequently smaller and less heavy at laying than first laid eggs Very few young died after about 20 days of age in peak vole years. The pattern of mortality in low vole years was markedly different. The majority of deaths still occurred in the first 20 days, but there was a substantial mortality of much older birds, even up to the point of fledging.

In Mali, mortality also occurred mostly in the first 20 days or so and was related to hatching order. The average age at death of first hatched chicks that perished was 18.2 days whereas for those that were sixth hatched or more the average age at death was only 5.6 days.

Young that died after they were about 2 weeks old in the Scottish area were often found lying among the general nest debris and in those cases where the carcass was fresh, it was clear that the chicks had been severely emaciated before dying. More often than not, young that died at an earlier age simply disappeared, but in most cases these chicks had lower than average weights when they were inspected at earlier visits, suggesting that they also died of food shortage. Occasionally, dead chicks were found in a partly consumed or dismembered state and, on a few occasions, their remains were found in the pellets of other chicks. The females probably treated dead chicks just as they would have other items of stored prey at the nest and fed them to the other chicks. This behaviour has also been recorded in France (Baudvin 1986) and presumably is of widespread occurrence.

Apart from starvation, there is little evidence of other significant causes of chick mortality. In Scotland, there was no evidence of predation on chicks but in France, Baudvin (1986) recorded apparent predation from beech martens (*Martes fiona*), although this only occurred at 2 out of 1031 nests that were inspected. The martens seemed to seek out similar nesting places to the barn owls. Racoons are known to be predators of barn owl nests in North America, but there are no estimates of the extent of this predation (Bruce Colvin, personal communication). There is also no information for tropical populations where, presumably, reptilian predators could be important.

All of the available information, certainly for temperate area barn owl populations, suggests that starvation is the most important cause of death among chicks and that most die at a relatively early age. The food demand of broods peaks at around 30 to 40 days, so most chick mortality occurs before this is reached. The highest mortality occurs while the female is still brooding the chicks and therefore not participating in hunting. The male must provide food for himself, the female and the chicks, and this may be the point at which food shortage becomes most critical.

In the last chapter it was suggested that the barn owl's small egg size would enable them to produce larger clutches but that this might present difficulties when rearing larger broods. By having the chicks hatch asynchronously, their number and the demands on the parents can be adjusted to the food conditions prevailing at the time. In a sense, barn owls keep their options open for as long as possible and this enables them to achieve a high productivity when food and weather conditions are favourable, and at least some output when conditions are poor.

12.4 Fledging success

The effect of food supply

The number of young reared by barn owls varies greatly in relation to many factors in the environment, but food supply seems to be the most important. In the Scottish population, just as laying dates and clutch sizes were correlated with yearly changes in vole abundance so also was the average number of young fledged. A cyclic pattern was clearly evident, which coincided exactly with the 3-year vole cycle (Figs. 12.8, 12.9). In peak vole years, when vole density was high throughout the whole of spring and summer, the owls produced on average around 4.0 to 5.0 young per pair from their first clutches. In the following years, when the voles declined in late winter and remained at low densities

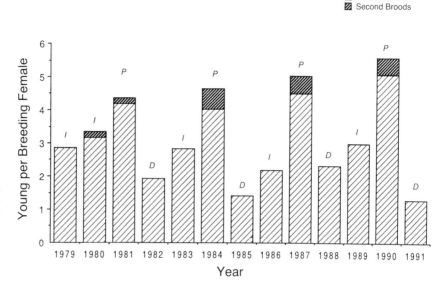

Fig. 12.8. Annual changes in the number of young fledged in relation to the field vole cycle in the Esk study area, south Scotland. *P*, peak vole year; *D*, declining vole numbers; *I*, increasing vole numbers.

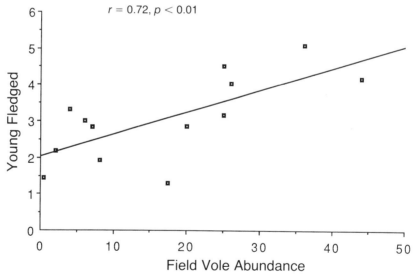

Fig. 12.9. Correlation between the average number of young fledged per pair each year from first broods and field vole abundance in the Esk study area, south Scotland.

throughout summer, they had the lowest productivities, at about 1.5 to 2.5 young per pair. In the third years of the cycle, the increase phase, when vole densities were low in spring but rose through summer and autumn, production increased consistently to around 3.0 young per pair. Second broods occurred only in peak years and involved a maximum of 15% of the females in those years and an average of only 4.5% over the whole study period. These added only a small amount to overall productivity (Fig. 12.8). The maximum number of young reared by individual pairs from first broods was eight and the maximum from pairs that reared two broods was nine.

Bigamy was recorded only seven times in the 14 years of this study when most breeding adults were trapped and ringed. Usually one female was clearly secondary and laid much later than the other. Such females were less successful, rearing on average only 1.1 young compared with 2.6 from the primary females. On two occasions, females were recorded who laid clutches with one male, remained until the young were about six weeks old, and then laid second clutches with new males on adjacent sites, while the first males successfully reared the initial broods.

Almost all European barn owl populations have shown large annual fluctuations in productivity. However, prey abundance was assessed quantitatively in only one other published study, in Germany, and this demonstrated a correlation between production and vole abundance similar to that found in Scotland (Schonfeld and Girbig 1975). The average number of young fledged from first broods in this study ranged from 2.4 per pair in the low vole years to 4.3 in peak years. This was substantially increased by the contribution from second broods in peak years which more than doubled overall production (Table 12.1). In another study, in Holland, common vole abundance was judged over 11 years by counting the numbers of fresh burrows (Braaksma and de Bruijn 1976, de Bruijn 1979). While this approach was not strictly quantitative, vole peaks and declines were obvious and followed a cyclic pattern with peaks every 3 to 4 years. The number of young fledged by the owls was once again closely correlated with vole abundance. Interestingly, although this population was some 500 km distant from the German population discussed above, annual variations in production in the two seem to have been in synchrony. A number of other studies have also shown fluctuations in production which were correlated with subjectively assessed changes in vole abundance and often synchronised across much of central Europe (Kaus 1977, Ziesemer 1980, Baudvin 1986, Chanson et al. 1988, Muller 1991). In years of peak production, individual pairs reared up to 10 or 11 young from first broods and some pairs succeeded in fledging as many as 16 young from two broods. Generally, pairs that reared two broods were those that laid early in the season and that had moderate-sized rather than exceptionally large first broods (Baudvin 1986, Muller 1991).

Seasonal changes in the average number of young fledged per brood followed the seasonal changes in clutch size noted earlier (Fig. 12.10 and Chapter 11). Thus in Scotland, there was a decrease as the season progressed whereas in many continental European populations more

Table 12.1. *Annual changes in the number of young fledged from two European barn owl populations in relation to common vole abundance*

					Year					
	1967	1968	1969	1970	1971	1972	1973	1974	1975	1976
Germany[a]										
Young fledged per pair										
First broods only	–	3.9	2.4	2.6	3.5	2.7	3.8	4.3		
First and second broods	–	3.9	2.4	3.7	7.1	2.9	3.8	7.0		
Females producing second broods (%)	–	12.0	0	37.5	64.8	17.1	0	55.6		
Vole cycle	–	1–2	1	2–3	3	1	1–2	2–3		
Holland										
Total young fledged per pair[b]	–	–	2.5	2.7	3.2	3.3	2.8	3.7		
Vole cycle[c]	3	2–3	1	1–2	3	2–3	1	3	2–3	1

Vole population: 1, decline year; 2, increase year; 3, peak year.
[a]Schonfeld and Girbig 1975.
[b]Braaksma and de Bruijn 1976.
[c]de Bruijn 1979.

Fig. 12.10. A young barn owl
of the *alba* subspecies almost
ready to leave the nest.
(Photograph: Iain R. Taylor.)

young were reared, on average, from clutches that were laid in June
and July than in April or May. In years of peak production more young
were fledged from second broods than from first broods laid some
3 months earlier (Schonfeld and Girbig 1975, Baudvin 1986, Muller
1991).

North American barn owl populations also show annual fluctuations
in breeding success, although somewhat less marked than those of European populations. However, in areas such as Utah and New Jersey
where microtine voles are the most important prey it is difficult to detect
any evidence of cyclic patterns in production. The Utah population
occupied an area of irrigated farmland where the owl's diet varied very
little from year-to-year and where the vole abundance was not cyclic
(Marti and Wagner 1985, Carl Marti, personal communication). The
number of young fledged each year varied around 3.6 to 4.8 per pair
except in 1982 when it fell to only 1.6 following an extraordinarily severe
winter which may have reduced vole populations. Production in the
New Jersey population ranged from 3.1 to 4.5 young per pair. Colvin
(1984 and personal communication) has shown that this variation was

caused by changes in the abundance of voles available to the owls in summer. In this case changes in vole abundance were brought about by the effect of spring rainfall on the growth of grasses upon which the voles depend. This area has hot summers with high evaporation and the growth of vegetation is thus limited by rainfall. Further south, in Texas, breeding success was sharply reduced from a normal range of about 2.0 to 3.0 to only 0.2 young per pair following a collapse in rodent populations. Only one pair out of 12 that attempted to breed in the year concerned actually succeeded in producing any young and, interestingly, this pair had managed to switch their diet almost entirely to redwinged blackbirds (Otteni *et al.* 1972). The only African population to be studied in detail showed a reduction in production associated with a decline in the abundance of multimammate mice (Wilson *et al.* 1986).

There is widespread evidence from many regions that barn owl breeding success is highly variable from year to year and at different times within years. All of these variations can be related to food supply both at the time of laying and later when feeding the young.

The effect of habitat

The influence of different habitats on barn owl breeding success has rarely been studied despite its potential importance, especially in conservation. If we consider habitat on a broad scale in Scotland, pairs nesting in young conifer plantations consistently reared more young than those nesting in the low farmland habitats. The difference was especially pronounced in vole peak years. Pairs nesting in the rough grazing habitat of sheepwalk were usually intermediate in their production (Fig. 12.11). These differences were related to the density of good quality vole habitat around the owl's nest site so that once again food supply was probably the important factor.

An interesting relationship was found between diet and breeding success among sheepwalk nesting pairs. Those with the higher percentage of field voles in their captures fledged more young. This was apparent in most years and was related mainly to the length of the grassland vegetation within the birds' hunting ranges. Where there were extensive patches of damp, lightly grazed rough grassland, voles dominated in the diet whereas on better quality soils where the rangeland had a higher stocking density and consequently shorter vegetation, common shrews dominated. As the voles were about three times heavier, on average, than the shrews, the owl's foraging efficiency was probably considerably reduced on the shorter grassland areas, leading to lower breeding success.

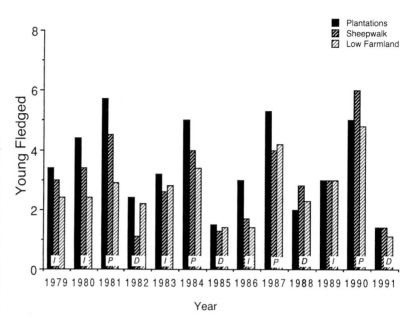

Fig. 12.11. Comparison of the mean number of young fledged each year by barn owl pairs nesting in plantations, sheepwalk and low, enclosed farmland habitats in the Esk study area, south Scotland. In most years, pairs nesting in plantation reared the largest broods and those nesting in low farmland reared the smallest broods. *P*, peak vole year; *D*, declining vole numbers; *I*, increasing vole numbers.

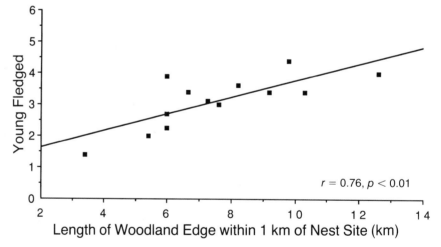

Fig. 12.12. The production of young from different nest sites, averaged over at least 8 years, in relation to the length of woodland edge habitat within a 1 km radius of the nest site. Data from the lowland, enclosed pastoral farmland section of the Esk study area, south Scotland. The greater the length of edge habitat, the higher was the owls' breeding success.

In the low pastoral farmland habitat the owls foraged mostly along woodland edges (Chapter 6). The densities of these foraging areas within the summer hunting range, on average a 1 km radius around the nest (Chapter 7), varied greatly from pair to pair and the long-term average production achieved at each nest site, assessed over at least 8 years, was closely correlated with these variations (Fig. 12.12). Nest sites surrounded by high densities of woodland edges were associated with higher productivities. There was considerable turnover at all of these sites; few individual owls survived longer than 3 years and there was no correlation between the frequency of occupation of various sites and the amount of woodland edge around them. Poorer quality ranges were not occupied consistently by younger birds. It is unlikely therefore that the relationship between productivity and the habitat around nests could be explained by the quality of the birds occupying them. The effect was probably a direct relationship with habitat, acting through differences in the density of available prey.

Edge habitats are often important foraging places in farmland (Chapter 6), so it is not surprising to find other instances where breeding success was related to the density of these edges around the barn owl nests. In Oklahoma, farmland barn owls did much of their hunting along roadsides where there were both fencerows to serve as perches and strips of habitat suitable for cotton rats, the main prey species (Goertz 1964, Ault 1971). Breeding success was followed at 11 nests over five years and a good correlation was obtained between the average number of young produced and the length of roadside within 1 km of each nest site.

12.5 Production in different populations

Populations of barn owls in different parts of the world vary considerably in their annual productivity per pair, in the amplitude of changes from year-to-year and in the contribution made by second broods. Long-term average production figures are available for a number of populations (Table 12.2). These range from 1.9 fledged young per pair in Mali and Texas to 5.3 in eastern France. The highest recorded values are from the prey-rich oil palm plantations of Malaysia, where most females are believed to produce two broods each year with a total annual production averaging around 7 young per pair (David Wells, personal communication).

Most central European populations produce about 4.0 to 5.5 young per pair, whereas in western Europe, in Holland and in Scotland, long-term averages are about 3.2 young per pair. There are no published results from long-term intensive studies in England or Wales, but evidence from other sources suggests a production of only around 2.5 to 3.0 young per pair during the 1980s (Percival 1990). These values, should they prove to be valid, are probably inadequate to maintain the population. Populations in the northern areas of the United States (New Jersey and Utah) produce about four young per pair each year. The published value for Utah over six years was actually three young per pair but this included an extremely unusual summer of very low productivity

Table 12.2. *Long-term average production figures for barn owl populations in different areas of distribution*

Locality	Duration of study	Average clutch size (n)	Average No. of young fledged per pair			Total production per pair per year	Author
			First broods	Second broods	Unspecified		
Scotland	13	4.6 (425)	3.1 (469)	3.05 (21)	–	3.2	Taylor, this study
France	8	6.0 (481)	3.55 (361)	5.02 (120)	–	5.3	Baudvin 1986
	14	?	4.0 (830)	4.7 (247)	–	5.04	Muller 1989, 1991
Germany	7	5.9 (309)	3.4 (231)	3.86 (80)	–	4.3	Schonfeld and Girbig 1975
	4	5.5 (354)	4.34 (158)	5.9 (69)	–	4.0–5.0	Kaus 1977
Holland	6	4.2 (477)	3.1 (452)	–	–	3.1	Braaksma and de Bruijn 1976
New Jersey	4	–	3.8 (125)	–	–	3.8	Colvin 1984
Utah	6	6.9 (146)	3.0 (146)	–	–	4.0	Marti and Wagner 1985
Texas	7	4.8 (88)	1.9 (88)	–	–	1.9	Otteni et al. 1972
Oklahoma	5	–	2.8 (32)	–	–	2.8	Ault 1971
Mali	4	6.05 (140)	1.84 (123)	1.9 (13)	–	1.9	Wilson et al. 1986
Malaysia	–	6.6 (36)	4.5 (15)	2.7 (14)	–	6.0–7.0?	Lenton 1984a
Galapagos Islands	–	3.1 (10)	–	–	1.6 (10)	1.6	de Groot 1983

following the exceptionally severe winter of 1979. Throughout the 1980s, including unpublished data, the figure is closer to 4 young per pair (Carl Marti, personal communication).

Some of these differences can be explained reasonably convincingly on the basis of food supply. The higher productivity of central European birds mainly arises from their extended breeding season and the increased frequency of second broods. Compared with western European populations, they depend more on common voles, whose availability to the owls is probably not reduced to the same extent as that of field voles by the annual growth in vegetation as the summer progresses (Chapter 5). The population in Malaysia, because of the greater stability of climate, had a prolonged season of high prey availability. Birds in areas with generally drier climates, in Texas, Oklahoma, Mali and on the Galapagos Islands, probably had lower overall prey densities and shorter periods of higher prey availability, associated with seasonal rains.

All populations that have been studied over a long enough period have shown some variation in production from year to year but the amplitude of these fluctuations differs among populations. Considering only first clutches, annual average production in the Scottish area ranged from 1.3 to 5.1 young per pair. Comparable ranges for some central European populations are from 3.0 to 5.1 (eastern France, Muller 1990), 1.9 to 4.3 (eastern France, Baudvin 1986) and 2.4 to 4.3 (Germany, Schonfeld and Girbig 1975). Therefore, the Scottish population experienced much more extensive variations. It is possible that prey abundance varied more in Scotland, but it also seems likely that diet diversity was important. Whereas the Scottish birds had only three important prey species, the central European birds had a much wider prey spectrum and consequently were able to turn to a greater range of prey when vole populations declined. The same general principle probably operated in the different habitats of the Scottish area. The variance of the annual average production figures over 13 years (a simple measure of the size of fluctuations around the average value) was 2.16 for the plantation habitat but only 1.15 for the low farmland habitat. In the latter, the owls could turn to both common shrews and wood mice when voles declined, whereas in the plantations, common shrews were the only alternative.

12.6 Summary

In most areas, about 70% to 85% of barn owl eggs hatch successfully. Values from North American studies are lower but these birds may be more susceptible to human disturbance and published figures may not be representative. When the chicks hatch, they are covered only in a thin layer of fine down and must be brooded by the female for the first 3 weeks or so. They grow rapidly, achieving half their final weight when about 18 days old. Skeleton growth is completed by about 40 days, but the primary feathers are not fully grown until about 70 to 75 days. The chicks hatch at about 2-day intervals, so that a considerable size difference is established among them from the start. In Scotland, only 5% to 10% of the chicks died in peak vole years but as many as 45% died in vole decline years. Most of this mortality fell on the last hatched chicks and occurred within the first 20 days of their life. Overall production in this population was cyclic corresponding to the 3-year vole cycle, with an average of 4.0 to 5.0 young raised per pair in vole peak years and 1.5 to 2.5 in vole decline years. Up to 15% of pairs reared second broods in high vole years. The rearing of second broods is more frequent in central Europe and can almost double average production in peak vole years.

13 *Dispersal*

The most interesting aspect of dispersal from the point of view of population dynamics and genetics is the distance the birds move between where they are hatched, their natal site, and where they eventually settle to breed. It is important to determine if local populations are self-sustaining, dependent mainly upon their own production or upon whether there is a greater degree of interchange and interdependence among populations. On a finer scale, it is interesting to discover the extent to which birds are faithful to the precise area in which they were raised. Dispersal can be important for the maintenance of genetic variation within populations of birds and particularly in the avoidance of close inbreeding by the spatial separation of near relatives. Inbreeding can result in increased homozygosity with the result that recessive deleterious alleles may become expressed, and this may lead to a reduction in reproductive performance (Greenwood 1987).

13.1 Comments on methodology

Dispersal patterns may be determined from national ringing schemes in which birds ringed as nestlings are subsequently found dead and reported by members of the public, or from intensive local studies in which the researchers ring nestlings and recapture them as breeding adults at their nesting places. The advantage of the second method is that certain evidence of breeding is obtained, but, if dispersal distances prove to be large, it becomes difficult to encompass a study area that is big enough to include all movements. National ringing schemes can give information for very extensive areas but will only give reliable results if the birds ringed are representative of the population at large and if those later found dead are representative of all birds that die. It is often particularly difficult to satisfy the second of these conditions.

The circumstances reported for recoveries of dead barn owls vary somewhat from country to country but two categories are almost always predominant: those found dead alongside roads and railways and those for which there is no information other than the date and place of death. Road and rail deaths can account for as many as 60% of all recoveries (Baudvin 1986, Percival 1990). This presents a potentially serious problem for the interpretation of recovery distances because some, and perhaps many, barn owls become stuck to the vehicles with which they collide and are transported considerable distances. In the first 2 years of my study I received six first-hand reports of this from truck drivers and I observed the phenomenon myself when a barn owl was struck by a coach in which I was travelling and was carried, attached to the wing mirror, about 200 km before being dislodged and thrown to the roadside.

To check if possible transportation had a significant effect on dispersal distances, I did a detailed analysis of the recovery of birds ringed in and around my study areas from 1978 to 1985. The finder's name and address is usually included on the recovery forms, so I contacted as many as possible to obtain details for those birds for which no information on recovery circumstances had been given. In this way I could compare samples of birds found dead on roads and railways with those that died elsewhere. The road-killed individuals had travelled a mean (geometric) distance of 44.2 km ($n = 23$) whereas for the others the mean distance was only 9.1 km. (Mann–Whitney U-test, $z = -3.86$, $p < 0.001$). The results may have been somewhat unusual as a major railway and an important four-lane highway carrying heavy traffic of large lorries cut through the area. However the general conclusion that the apparent dispersal distances of road-killed owls was greater than that of those that died in other circumstances was later substantiated in an analysis of all British barn owl recoveries up to 1988 (Percival 1990). There may be other explanations for these results, but transportation must at least be a contributory factor. Ringing recoveries therefore probably tend to exaggerate average dispersal distances and long-distance recoveries must be treated with some scepticism unless precise details of the cause of death are given. Considerable care must be exercised when interpreting recoveries and precise values should not be accepted uncritically. However, their use to make general comparisons, especially of year-to-year variations within the same region and of different regions showing similar recovery circumstances, is almost certainly valid.

Data are available from national ringing recoveries for a variety of populations in Europe and North America. However, results have been expressed in a wide variety of forms which makes precise comparison difficult. The most useful approach would be to present histograms showing the numbers and densities of recoveries of breeding-age birds at various distances from their natal sites. For genetic analysis it is useful to have a very fine resolution in the shorter distance categories. Some authors have presented only mean values, sometimes arithmetic and sometimes geometric; distances have been given in miles or in kilometres and a variety of distance categories has been used. More often than not, all recoveries have been lumped together irrespective of the age of the birds. Without the original data, standardisation is impossible but there is obviously a good case for some reanalysis. I would suggest 10 km distance categories up to 50 km and 50 km categories thereafter, with a much more detailed analysis of recoveries within the first 20 or 30 km and a clear separation of age classes.

13.2 Post-fledging dispersal in Europe

In the Scottish study areas, where many of the birds nested in abandoned farms, I often found youngsters still roosting in buildings adjacent to the nest some two to three weeks after they had fledged. Parental feeding was sometimes seen, especially in the first one or two weeks. However the young usually disappeared soon after this, suggesting that they had begun to disperse away from their home areas (Fig. 13.1). A much more thorough investigation of the age at which young start to disperse was done on the island of Anglesey, in Wales, by attaching back-pack radio transmitters to a sample of 10 newly fledged birds (Seel *et al.* 1983). Two of the radios failed but of the remaining eight birds, two moved away from the nest area 2 to 3 weeks after fledging, two at 3 to 4 weeks, two at 4 to 5 weeks and two between 6 and 8 weeks. Most obviously started their dispersal phase very soon after leaving the nest but considerable variation among individuals was also evident, although no explanation of this was possible.

By 1 month after ringing, and hence presumably within the first week or two after fledging, 37% of British recoveries (up to 1978) were at

Fig. 13.1. A young barn owl makes a landing after its first flight. (Photograph: Robert T. Smith.)

distances greater than 3 km from the nest. At around 5 or 6 weeks after fledging this had risen to 59% (Bunn *et al*. 1982). In a later analysis, Percival (1990) estimated that the mean (geometric) distance of recoveries up to about 12 weeks from ringing was around 3 to 4 km from the nest. So, although most young birds moved away from the nest soon after fledging, it seems that the majority did not travel all that far.

By contrast, in Germany, 27% of recoveries within the first month after ringing as nestlings were at distances greater than 10 km from the nest and 18% were between 50 and 100 km. In the second month, only 27% were found less than 10 km away and almost 40% were more than 50 km from their nest (Bairlien 1985). Again, great variation was apparent among individuals with some moving a long way and some remaining quite close to home. Nevertheless, it seems that most of these continental European birds dispersed at an earlier age and to much greater distances than did the British birds.

Dispersal distances stabilised in the German population about 4 to 5 months after fledging. The birds may have continued to move around after this time, without changes in overall distances, but it seems likely that the most significant part of their post-fledging dispersal was completed within these first 4 months and before the start of winter.

Rapid, early dispersal may be advantageous if there is likely to be competition for suitable vacant nest sites in good quality habitat or if there are possibilities of pairing with older, more experienced unmated birds. Indeed, some birds in the Scottish study area definitely settled at their subsequent breeding places within 2 months of fledging. One case involved a female who was found roosting alongside a single male at a somewhat isolated nest site within 5 weeks of fledging. The male was in his fourth year, occupied good quality habitat and had bred successfully as a 1 year old but had failed to attract a mate in the following two springs. His nest site was littered with freshly killed prey and observation revealed that he was very actively providing food to his new female. She even occupied the nest chamber for a week or two but did not lay. However the pair bred successfully in the subsequent 2 years.

Remaining within the parental home range could also be disadvantageous to both offspring and parents as this might result in a very high density of foraging birds with competition among them: brood sizes are not infrequently as high as six to eight. On the other hand, if the parental range and the immediately surrounding area is very good quality habitat with high densities of prey, it may be advantageous for the young to remain at least until their foraging skills are reasonably well developed. The extent of parental post-fledging feeding might also be important in deciding when and how the young move and when two broods are reared this period is presumably reduced. The relative importance of all these factors at different times and places might explain some of the variation in dispersal behaviour found among individual youngsters and this would be a fruitful area for further more detailed research involving radio-tagged birds.

Few studies based on ringing returns have given information specifically concerning birds that have reached breeding age and that presumably have settled on their breeding areas. In those that have done so, estimated dispersal distances are only very slightly greater for such birds than for all birds lumped together regardless of age. This again supports the earlier evidence that the birds undertake most of their dispersal very soon after leaving the nest. When examining dispersal data to obtain information on the distances travelled from natal to breeding sites, studies that do not subdivide age categories can therefore be used alongside those that do.

As was found for the very early stages of dispersal, there is considerable variation within and among different European populations in the distances from natal to presumed breeding sites. Some birds moved only a few kilometres, whereas others seemingly travelled hundreds of kilometres. Some from the French population studied by Hugues Baudvin (1986) were recovered in the south of Spain at distances of up to 1200 km and even greater movements of about 1600 km have been reported from Holland and Germany to Spain (Braaksma and de Bruijn

Table 13.1. Dispersal distances obtained from national ringing scheme recoveries for a number of European barn owl populations

Country	Sample size	Recoveries (% of total) at varying distance from natal site (km)						
		0–10	11–20	21–30	31–40	41–50	51–100	>100
Germany[a]								
Under 1 year old	878	←		45.9		→	32.0	22.1
Over 1 year old	367	←		37.6		→	32.2	30.1
Britain[b]								
Under 1 year old	184 }	61.5	←	30.9		→	5.3	2.3
Over 1 year old		53.3	←	37.0		→	4.9	4.8
France[c]								
All ages	316	←	48.4	→	←	23.4 →	8.9	19.3
Holland[d]								
All ages	793	←	53.3	→	← 19.4	→	12.5	14.7
Scandinavia[e]								
All ages	143	34.3	21.7	14.0	7.0	9.1	18.9	9.1

[a]Bairilien 1985;
[b]Bunn et al. 1982;
[c]Baudvin 1986;
[d]de Jong 1983;
[e]Frylestam 1972.

1976, Glutz and Schwarzenbach 1979). Sauter (1956) gave details of two young of the same brood, one of which was recovered in Spain at 1380 km, the other in the Soviet Union at 1260 km.

British barn owls seem to disperse over much shorter distances than birds from any of the continental European populations that have been studied (Table 13.1). About 53% of breeding-age birds in Britain were found within 10 km of their natal site and 90% within 50 km. Less than 5% had gone more than 100 km. Bearing in mind the exaggeration of distances arising from the high percentage of road deaths in recoveries, it is likely that the great majority of British birds settle to breed within 30 to 40 km of their birth places. Percival (1990) estimated that the mean (geometric) distance travelled by breeding-age birds was only about 8 to 12 km. In comparison, less than 40% of the adult birds recovered in the German populations studied by Bairlien (1985) were within 50 km of their natal sites and 30% were found at distances greater than 100 km. The Scandinavian population (mostly in Denmark) was closer to the British population than the others with 86% within 50 km (Frylestam 1972) and the Dutch and French populations were inter- mediate between these and the German birds with about 70% of recov- eries within 50 km (Bunn *et al.* 1982, Baudvin 1986). It is unlikely that differences of such magnitude could be artefacts of recovery circum- stances and it must be concluded that they have some fundamental ecological significance.

13.3 Comparison of the sexes

Recoveries from national ringing schemes cannot be used to compare males and females as the sexes are not differentiated either at ringing or on recovery. This problem can be overcome in intensive studies where the young are ringed as chicks and subsequently recaptured on their breeding sites where their sex can be determined. Only females develop brood patches so there can be no confusion. However, for this method to work, study areas must be large enough to encompass most movements, otherwise a new bias will be introduced. Clearly for popu- lations with very large dispersal distances such as the German one described earlier, this approach is not practical but it can be applied to British birds which have much more limited dispersal. Of course, unqualified extrapolation from one population to another should not be made.

In my main Scottish study area, I captured breeding adults from 1980 to 1991, resulting in the retrapping of 47 females and 36 males that I had ringed as nestlings. Dispersal distances from hatching to nesting

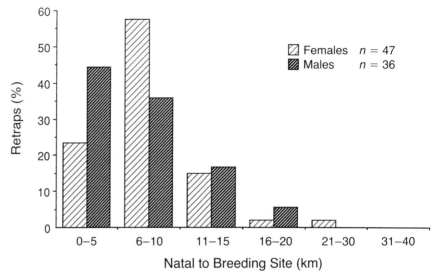

Fig. 13.2. Distances from natal sites to breeding sites of male and female barn owls in the Esk study area, south Scotland. Females moved significantly farther than males (Mann–Whitney U-test, $z = -1.81$, $p = 0.035$).

places for this population were very short and only a single individual was recovered more than 20 km from its natal site. The density of retraps declined very rapidly away from the natal sites: only 17% of the birds settled more than 10 km and 4.8% more than 15 km away from their hatching places (Fig. 13.2). These distances were not biased by the size of the study area which measured 1600 km² and the birds could have travelled further had they simply dispersed randomly within the area. The short distances are probably explained to some extent by the relatively good quality habitat within the area and the availability of closely spaced suitable nest sites.

The difference between males and females was not great but nevertheless males settled to breed significantly closer to their natal sites than did females. Males settled most frequently within 5 km of their birth places whereas females most often settled between 6 and 10 km away (Fig. 13.2). Similar sex differences, with females moving further, have been recorded in a few other birds of prey, for example sparrowhawks (Newton 1986) and hobbies (Fuiczynski 1978).

13.4 Direction of dispersal in Europe

Birds in the Scottish study area showed no directional bias in their dispersal and there was also no bias in a sample of 400 recoveries for the whole of Britain (Bunn *et al.* 1982). In a more detailed study of German recoveries, birds were found to disperse in all directions but there was a definite tendency for more to move southwards, particularly

in a south-west direction. This tendency was somewhat more pronounced in movements greater than 50 km from the nest sites compared
with those over shorter distances and was established within 2 to 3
months after fledging (Bairlien 1985). A southerly and westerly bias,
especially in these longer movements, seems to occur widely throughout
continental Europe and has been recorded for birds from Holland,
France and Switzerland as well as from Germany (Sauter 1956,
Schonfeld 1974, de Jong 1983, Juillard and Beuret 1983, Baudvin 1986).
Many fewer recoveries have come from a south-easterly direction and
it has been suggested that the Alps form a barrier to dispersal in that
direction.

There is no evidence of a return movement of these birds back
towards their natal areas in the following spring. Of German birds
recovered over 50 km from their natal sites 9 to 12 months after fledging,
when they should have settled to breed, 48% were still in a southerly
direction. Also, 36.5% of recoveries between 9 and 12 months were
still more than 100 km from their natal sites (calculated from data in
Bairlien 1985). Thus it seems likely that many of the birds that moved
southwards remained there to breed. This implies a general 'leakage'
of production from northern and central European populations, particularly into France and suggests a much greater amount of interdependence and genetic interchange among these populations than is apparent
in Britain.

13.5 Dispersal of North American barn owls

So far, the discussion of natal dispersal has concerned only European
populations. Rather less information has been published for North
American barn owls, although extensive data sets remain to be analysed
and published.

There seems to be a marked difference between birds in the south
(up to 3° north) of the United States and those in the north (Stewart
1952). Most recoveries of southern birds (mostly in California) were
within 80 km (50 miles) of their hatching places (88%) and none were
more than 160 km (100 miles) ($n = 67$). By contrast, only 56.8% of
169 recoveries of birds ringed in northern states (particularly Ohio,
Pennsylvania, New Jersey and Massachusetts) were recovered within
80 km (50 miles) of their natal sites, and 31% were found at distances
greater than 160 km (100 miles). Considering only individuals more
than 6 months old, 43.7% of the northern birds were found more than
160 km (100 miles) and 27.7% more than 320 km (200 miles) from
their hatching places. The direction of movement of these longer dis-

placements was predominantly southerly, with individual birds reaching Florida, Georgia, Alabama and Mississippi, and this prompted Stewart to conclude that the north-east populations were partly migratory. However the case for this is far from conclusive, at least on the basis of currently published work. Stewart found that there were no recoveries of birds more than 320 km south of their natal areas during the summer months, and cited this as evidence for a northward return movement. However, few deaths and few recoveries would be expected in the summer months, even in a sedentary population (Chapter 14) and the number available to die would also be minimal at this time, having been reduced by cumulative mortality throughout the winter.

Sitings of exceptionally large numbers of barn owls in the autumn months at some east coast observation points such as Cape May Point (New Jersey) are also interpreted in support of a southward migration, but the appearance of such birds could simply be a result of autumn dispersal, with no specific direction, of juveniles from adjacent areas of good habitat.

The long-distance movements could possibly arise in a similar way to comparable movements in Europe, as a response to periods of low prey abundance (see below). This might then explain some differences apparent within the east coast populations that Stewart suggested comprised some migratory and some non-migratory birds, a phenomenon which is difficult to explain otherwise. A convincing resolution of this question must await the publication of further information.

Some of the long-distance recoveries in North America raise other interesting questions. A bird ringed as a nestling in New Jersey in 1981 was recovered in Bermuda 20 months later, a cross-water journey of 1250 km (Soucy 1985). Did the bird make the trip unaided or was it transported on board ship? Mueller and Berger (1959) reported a barn owl that had been ringed as an adult in Wisconsin and subsequently captured on a ship 360 km east of Savanna, Georgia, so either explanation may be possible. A few European barn owls have been discovered on North Sea oil rigs in circumstances that suggest they had arrived unaided (A. Anderson, personal communication).

13.6 Dispersal distance in relation to vole abundance

It is well known that the distance that young continental European barn owls disperse from their nests varies greatly from year to year. Years of increased dispersal have been termed 'Wanderjahren'. These involve mainly newly fledged birds, although some adults also seem to be included, and are unlikely to be a response to weather as they are substantially completed before the onset of possibly adverse winter conditions. Honer (1963) and Sauter (1956) suggested that they were correlated with low or declining vole populations but could not supply quantitative evidence. However, by drawing together data from a variety of sources, it is possible to provide reasonably convincing circumstantial evidence in support of their idea. For the years 1971–77, four sets of pertinent data coincide (Fig. 13.3). Firstly, dispersal distances from national ringing recoveries are available for Holland (Bunn *et al.* 1982) and adjacent areas of Germany (Bairlien 1985). In 1971, 1974 and 1977, young birds from both these areas dispersed over much shorter distances than in the intervening years. Next, in Holland, the number of dead barn owls processed by taxidermists was significantly lower in these 3 years than in the others, suggesting that mortality was low and, circumstantially, that food supply was high (de Jong 1983). Survival in the German population, estimated by the percentage of ringed birds reaching their second year, was also high in these years. Studies in Holland, in the province of Groningen, showed that common vole abundance was indeed high in 1971, 1974 and 1977 (an exact 3-year cycle) and low in the other years (Wijnandts 1984).

Thus it seems likely that young barn owls fledged in years of high vole populations are able to find areas of good food supply relatively close to their natal sites and do not disperse over great distances, whereas those hatched at times of poorer food supply are forced to travel further in search of suitable conditions (*Wanderjahren*). Presumably, since the good years follow poor years during which the population declines, there will also be greater numbers of suitable, vacant nest sites closer at hand. The greatly increased mortality rates in the poor food years suggests

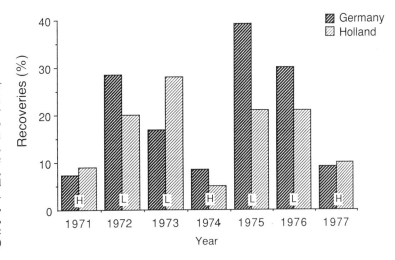

Fig. 13.3. The percentage of yearling barn owls recovered more than 100 km from their birth places in relation to common vole abundance in Holland and Germany. H, high vole year; L, low vole year. When voles were scarce, more of the owls dispersed over greater distances. (Sources. Owl data: Holland, Bunn *et al.* 1982; Germany, Bairlien 1985. Vole data: Wijnandts 1984.)

that, for most of the young birds, the long search for areas of higher prey abundance goes unrewarded. However, for this behaviour to have evolved, the advantages of dispersing must still outweigh those of remaining.

I could find no difference between high and low vole years in the distances moved by the Scottish barn owls and, throughout the whole of Britain, dispersal distances do not differ from one year to the next (Percival 1990). Presumably, in Britain, the disadvantages of remaining are not exceeded by the gains from dispersing farther.

13.7 Fidelity to nest site and mate

Reliable estimates of nest site and mate fidelity can be made only by catching and marking a high proportion of the breeding birds in a study population each year over several years. Studies that have not tried to find all nests and have involved the trapping of only a small proportion of the adults could give highly biased results. Overall, I was able to catch, on average, almost 90% of the breeding females and 75% of the breeding males each year in the Scottish area. Additional information was obtained for several birds by comparing moulted feathers found around nest sites with photographs of the extended wing taken in previous years. No two birds have exactly the same pattern of markings (Fig. 8.7, p. 118).

Over 13 years of study I caught 137 males and 150 females in each of 2 consecutive years. All but one of the males had remained on the same nest site from one year to the next, giving a site fidelity of 99.3%. Seven of the females moved to new nest sites, giving a fidelity of 95.1%. The single male that moved did so to an adjacent nest site following the loss of his mate during the winter and five of the females also moved after the loss of their mates. Two females moved after breeding for 1 year as the second female of bigamous males who did not die and one bird changed her nest site and mate twice in 3 years. Movements were all less than 8 km and in all but one case were simply to an adjacent nest site where they paired with males who had also lost their mates during the winter.

The movements of adults determined from national ringing recoveries in Britain were also negligible, averaging only in the order of 4 to 6 km (Percival 1990) which indicates that most died within their normal winter home range and had not shifted breeding places. The relatively sedentary nature of British barn owls, earlier described for juveniles, is also confirmed for breeding adults.

There is rather little comparable information from continental European populations, although several studies are in progress. Baudvin (1986) retrapped 34 birds in successive breeding seasons including eight males which were all recovered at the same nest and 26 females of which only one had moved, to a site 2.5 km away.

Evidence from national ringing schemes in Germany and Holland suggest that adult barn owls may be more mobile between breeding seasons in these countries. In Germany, 50% of 144 birds ringed as breeding-age adults were subsequently found more than 50 km from their place of ringing (Bairlien 1985) and 26 (55.3%) of 47 Dutch adults were recovered more than 10 km from their ringing place (Bunn *et al.* 1982). There must be a possibility though that some of these may not actually have been breeding at the time of ringing.

North American birds seem also to vary in their site fidelity. In the New Jersey population, seven (18.9%) of 37 males and 19 (38%) of 50 females trapped in consecutive breeding seasons had moved to new nest sites (Colvin and Hegdal 1986, 1987, 1988, 1989, 1990). However, most of the adults in the Utah population studied by Carl Marti (personal communication) were faithful to their sites from year to year.

The significance of these differences among populations is not obvious. By remaining at a particular nest site, the birds presumably gain by learning the details of local foraging conditions and have the security of a known nest site, but may suffer increased mortality when prey abundance declines. Moving might result in the discovery of better food conditions but entails the risk of failing to find a vacant nest site. It is possible also that site fidelity is lower in areas such as New Jersey that have a higher risk of nest predation from mammals such as racoons. Retaining the same mate each year could result in better breeding success unless a previous relationship was unsuccessful. The selective gains to the birds in staying or moving presumably depend upon the balance in any area of these advantages and disadvantages. The Scottish birds moved mainly when they lost a mate but, at other times, pairs often had the opportunity to shift to better quality ranges within 2 or 3 km, but did not do so. Possibly the advantage of experience on their own range outweighed any immediate gain from better quality habitat.

13.8 Dispersal and the avoidance of inbreeding

The discussion of dispersal has so far concentrated upon its relationship with environmental variables: temporal and spatial changes in food supply and possible competition for resources such as food or nest sites. However, it has frequently been suggested that one of the main causes for the evolution of dispersal behaviour is the avoidance of inbreeding

(Johnson and Gaines 1990). This, of course, implies that dispersal has a genetic basis and, although this might seem a reasonable assumption, there is as yet very little evidence for it. The technical problems involved in obtaining such evidence are considerable.

In the Scottish area, breeding males and females were highly faithful to their nest sites and mates from one year to the next. Thus parents could only have bred with their offspring if the young birds had settled on the parental home range. Likewise, the chances of full sibs (brothers and sisters) breeding together would depend upon their relative dispersal distances. Therefore, I reanalysed the data on dispersal distances presented in Fig. 13.2, but this time, instead of simply measuring linear distances, I calculated the number of home ranges between the birds' hatching places and their subsequent breeding places. This was done by drawing a 1 km radius circle (the usual summer range) around each site that had been used for nesting and counting the number of these circles transected by a line joining the birds' natal and breeding sites. This approach is not completely satisfactory but at least it gives a better impression of spatial relationships.

As expected, females tended to move over significantly more ranges than males before settling, but in both sexes about 90% settled within the surprisingly short distance of three home ranges of their natal sites. More males (36%) than females (15%) bred on the range immediately adjacent to that of their parents, their own natal site, but none was ever found breeding on the natal site itself (Fig. 13.4). This alone suggests that parent–offspring matings were unlikely to occur. Of course, it is possible that parents and offspring simply recognised and avoided each other. However, on six occasions when the young bred on an adjacent

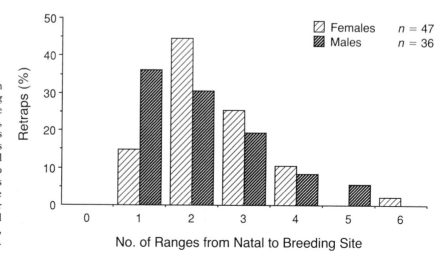

Fig. 13.4. Distances from natal sites to breeding sites of male and female barn owls, Esk study area, south Scotland expressed as the number of home ranges between the sites. 0, natal range; 1, range adjacent to natal range, etc. Females moved over significantly more ranges than males and neither sex settled on their natal sites (Mann–Whitney U-test, $z = -1.84$, $p = 0.033$).

range, both parents had disappeared, presumed dead, and were no longer on the natal range. In four of these cases, the natal range remained unoccupied throughout the summer. There are two possible explanations: either the birds avoided their natal range, irrespective of the presence of their parents, or they avoided their parents and settled permanently at their future breeding range very soon after fledging. Some evidence for early dispersal and settlement has already been presented. However, in seven of the 13 cases in which males settled on sites adjacent to their birthplaces, these sites were already occupied by a breeding pair at the time the young males fledged and started to disperse. They took possession of the nest sites following the death, some time over winter, of the former resident males. Unfortunately nothing is known of the movements of the young males during those winters; they may have wandered more widely to detect any vacancies in the general area or they may have been tolerated overwinter within the range in which they settled. The lack of strict territorial behaviour of resident adults, which was described earlier (Chapter 7), would give the youngsters considerable freedom.

The survival rate of young birds to breeding-age averages around 30% and even fewer manage to recruit to the breeding population (Chapters 14 and 15). Therefore, in the great majority of cases only one or two young from each brood manage to breed. Given that they tend to disperse in random directions from the nest and that females tend to go slightly further than males, it is unlikely that brothers and sisters would ever interbreed. I was able to establish the identities of both birds in 157 pairs examined between 1980 and 1991 and there was no case of parent–offspring mating and only one of full sibling mating. These were from broods of different years and presumably the birds were unable to recognise each other as brother and sister. Thus, the incidence of very close inbreeding is extremely rare. Nevertheless, the very short dispersal distances of both males and females could mean that a high proportion of birds in an area are quite closely related to each other.

13.9 Summary

Most young barn owls move away from their birth places soon after fledging and have completed their dispersal phase within about 3 months. In Britain, the majority settle to breed within 10 km of their hatching places and few move more than 50 km. By contrast, continental European barn owls disperse much further: in Germany most recoveries of breeding-age birds are more than 50 km from their hatching sites. In

North America, birds from the southern states disperse over shorter distances than those from northern states. Many of the latter have been recovered as far south as Mississippi, prompting suggestions of migration but there is little good evidence for this. There is no conclusive evidence of regular migrations for any population although birds in some countries, such as Australia, may be partly nomadic.

In Britain, dispersal distances do not change with annual fluctuations in food supply but in continental Europe years of low vole populations are associated with substantially increased dispersal and have been termed *Wanderjahren*. Mortality rates are exceptionally high during such years.

In Scotland adult barn owls are highly faithful to their nest sites and mates from one season to the next. This seems also to be the case in France and Utah (USA) but much greater mobility has been observed in some populations, such as those of the eastern United States. Male barn owls settle significantly closer to their birth places than do females and both tend to disperse in random directions, or at least independently of others in the same brood. None have been observed to breed on their natal sites. Thus, close inbreeding between parents and offspring and between brothers and sisters is rare.

14 *Mortality*

Mortality is one of the most difficult aspects of barn owl ecology to study and quantify as only a small percentage of dead birds are ever found, even when intensive searches are undertaken. Most published estimates of mortality rates are calculated using indirect methods based on the returns from national ringing schemes. However, the use of such ringing recoveries is only valid if it can be demonstrated that the birds found dead and reported are representative of the population as a whole. It is particularly important that the reporting rate be independent of the age of the birds, the cause of death and the location; when recovery circumstances are unusual, these conditions may not be met. A frequently quoted example illustrating this concerns goshawks in northern Europe where ringing returns were highly biased towards those birds killed by man, a fate to which young birds were more vulnerable than older birds. Thus mortality rates, especially for young birds were exaggerated (Haukioja and Haukioja 1971).

Most ringing recoveries for barn owls are of those killed on roads and railways. Many are returned with no details of recovery circumstances but a proportion of these are also road deaths. The owls often roost in inaccessible and rarely visited places such as holes in trees, down chimneys and in the lofts of disused buildings and one would expect that those that die of natural causes such as starvation, would often do so at these roosts and thus remain undiscovered by members of the public. On the other hand, road deaths are conspicuous and so one would expect them to be over-represented. One might also expect young, inexperienced birds to be more vulnerable to collisions than older birds and this has been shown to be the case in France where 73.2% of the recoveries of first year birds were road and rail deaths whereas the equivalent figure for older birds was only 53.4% (Baudvin 1986). An unrepresentative recovery rate biased towards birds killed on roads has been reported from Germany (Illner 1992).

In the Scottish area, between 1980 and 1986, I carried out an extensive programme to recover dead owls by asking farmers, gamekeepers, local natural historians and wildlife rangers to look out for them as they went about their work and as they travelled around the area. At the same time I carried out monthly systematic searches of likely roosting places and also drove along most of the roads checking for carcasses. This yielded 138 dead first year birds representing 19.8% of all young raised in the area and 66 older birds. Of the older birds, 43 (65.2%) were found in an emaciated condition at known roosts and only 15 (22.7%) were road deaths. By contrast, 78 (56.5%) of the young birds were found dead by roads, and 46 (33.3%) at roosts. This probably underestimated the number of younger birds dead at roosts as they often roosted in less predictable places than the older birds. National ringing returns covering the whole of Britain suggest that 49% of first year and 48% of older birds died on roads between 1983 and 1985 (Percival 1990). The intensive study approach was not without its own biases and sample sizes were smaller, but the view that national ringing recoveries were not representative and that the deaths of older birds may have been under-represented is given some support. Ringing recoveries probably over-estimate first year mortality in comparison with older age groups and also misrepresent the causes of death by over-emphasising road deaths, so published figures must be treated with caution.

In Britain, the proportion of first year and adult recoveries attributed to road deaths remained unchanged from 1970 to 1988 (Percival 1990) so the use of recoveries to compare mortality rates over time might seem to be valid. However, there has been an increasing tendency for more young birds to be ringed in specific conservation areas. Since these are often chosen because of their high densities of barn owls, it is likely that they are areas of relatively good habitat. By the provision of nest boxes and contact with farmers, breeding places are given greater security. The national ringing recoveries, therefore, may be becoming less and less representative of barn owl populations over wider areas and this could lead to difficulties in the analysis and interpretation of recoveries. For example, an increase in survival rates over time might be indicated whereas what might actually be happening is a gradual increase in the proportion of birds ringed in these patches of better quality habitat. Clearly, data from ringing returns must be analysed with great care and with a detailed knowledge of the circumstances.

Another method of quantifying mortality rates, which is only suitable for intensive studies, involves the disappearance rates of previously known ringed birds. Such birds may either have emigrated or died and if emigration can either be discounted or quantified, their disappearance rates can be used to estimate mortality. This is the method I used in the Scottish area to quantify the mortality rates of breeding adults. It was shown earlier (Chapter 13) that these birds were almost entirely faithful to their nest sites and those that changed sites had only moved a few kilometres. Also, all of 54 ringed adults found dead by myself and members of the public were within 6 km of their nest sites which was within their normal winter range (Fig. 14.1). The recovery rate for barn owls is so high that had significant numbers travelled greater dis-

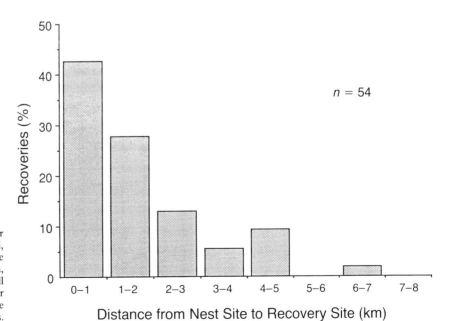

Fig. 14.1. Distance from their nest at which dead, ringed, previous breeding adults were discovered. Esk study area, south Scotland. Almost all died within their usual winter home range with no evidence of movements.

tances many would have been found and reported. Thus emigration could be eliminated as a possible cause of disappearance. The birds were highly faithful to their roosting places as well as their nest sites so their disappearance could easily be detected and those that did so were assumed to have died. No bird classified as dead in this way in a particular year was ever found alive or reported dead in subsequent years.

14.1 Seasonal patterns of mortality

Mortality was strongly seasonal in the Scottish area (Fig. 14.2). First year birds experienced a high death rate in their first autumn, peaking in October and November, and again during mid to late winter. Most were found dead in January, with reducing numbers through to April. Fewer were found in February than in autumn, but, because the numbers still alive and hence at risk was also less by that time, the mortality rate, as a percentage of those alive, was actually higher. Adult losses were greatest from December to March, with the highest mortality rates in January and February. This pattern of deaths seems to be common to all barn owl populations in temperate parts of Europe and North America (Stewart 1952, Schifferli 1957, Verheyen 1969, Frylestam 1972, Glue 1973, Braaksma and de Bruijn 1976, de Jong 1983, Juillard and Beuret 1983, Joveniaux and Durand 1984, Bairlien 1985, Baudvin 1986, Newton, Wyllie and Asher 1991). A more southerly population studied in central Spain showed a high autumn mortality but a less pronounced winter mortality (Fajardo 1990).

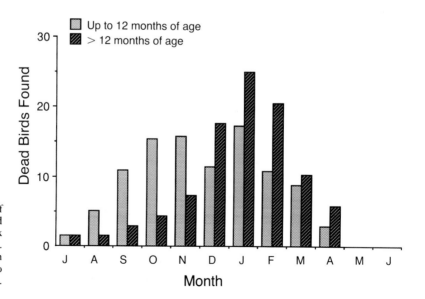

Fig. 14.2. Seasonal pattern of the deaths of first year and older barn owls in the Esk study area, south Scotland. Most mortality occurred in winter, from December to March.

The autumn deaths of young birds in Scotland occurred not immediately after leaving the nest, which for most birds was in July, but some 2 to 4 months later suggesting that it was not just the lack of experience or immediate post-fledgling dispersal into unknown habitat that was responsible. Prey abundance was still increasing during this period and most growth of vegetation had already occurred, so access to prey was probably not declining. It is possible that the gradually worsening weather conditions, with falling temperatures and increasing rainfall and windspeed could have been important, by lowering prey activity and reducing the young birds' hunting efficiency whilst at the same time increasing their energy requirements (Chapter 6). Many young were above adult weight at the time of fledging (Chapter 12) so perhaps their reserves offset their inexperience in hunting for some time so that deaths occurred later.

The high mortality rates of both first years and adults in winter points to the possible importance of winter weather conditions. This will be considered in more detail later.

14.2 Mortality rates in relation to age

Most birds of prey experience high mortality rates during their first year, for a variety of reasons, and lower mortality rates in subsequent years. Small species, roughly similar in size to the barn owl, such as European kestrels and sparrowhawks, commonly have first year mortalities of about 50–70%, falling to around 30–40% in subsequent years (Newton 1986, Village 1990).

I have calculated age-specific mortality rates, using Haldane's (1955) method, from published ringing recoveries of barn owls in a number of European and North American populations and from unpublished recovery data for south Scotland (Fig. 14.3). More sophisticated statistical methods of analysis are available (see Lebreton *et al.* 1992) and although these may allow greater precision, they would not fundamentally alter the general picture presented here.

All barn owl populations in temperate climates show remarkably similar patterns, with first year mortality rates of 65–75%, second year rates of around 40–60% and third year rates of 30–40%. Thereafter sample sizes are rather small for reliable analysis but there is a suggestion of a further reduction in the fourth year and perhaps an increase in subsequent years, although this needs to be confirmed by more detailed analysis of larger data sets. Temperate area barn owls thus experience high mortality rates throughout their lives but, interestingly, these seem to be only slightly greater than those of other small birds of prey in the

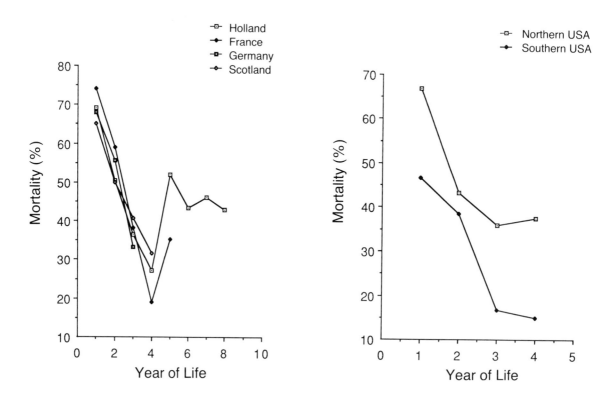

Fig. 14.3. Age-specific
mortality rates of barn owls in
Europe and North America.
Sources: south Scotland,
Taylor, this study; France,
Baudvin 1986; Germany,
Bairlien 1985; Holland, de
Jong 1983; USA, Henny
1969.
same areas. It has often been suggested that barn owls have very high
mortality rates in comparison with this group. Generally, only about 6
to 12% of all barn owls that fledge survive beyond 3 years of age. The
oldest bird in the Scottish area lived to be 11 years old but there are
records of individuals surviving to 18 years (Henny 1969, Braaksma and
de Bruijn 1976) and even apparently to 34 years (Keran 1981). The
average expectation of further life on leaving the nest has been estimated
to be 1.3 years in central Europe (Schifferli 1957) and 1.1 years in
north-eastern USA (Stewart 1952).

The higher mortality rate of first year birds probably arises from their
general lack of experience and the time taken to develop the skills of
locating and catching prey, although no studies of this have been made.
There is evidence that they are not excluded from good quality foraging
habitat by the territorial behaviour of adults (Chapter 7) which is a
common cause of increased mortality among immatures of other species
such as the tawny owl (Southern 1954, Hirons 1976). Why mortality
rates should continue to decrease, perhaps even up to the bird's fourth
year, is much less clear. Continued improvement of hunting skills is
possible but seems unlikely. It may be that only those that occupy the

best quality habitat survive to a good age, so that the decreasing annual mortality rates in older birds is mainly a habitat effect.

Barn owls living in more southerly latitudes seem to have significantly lower mortality rates (Fig. 14.3). The only published information is from the southern USA, mostly California, where first year mortality was 46.7% falling to 38.5% in the second and 16.7% in the third year. The average life expectancy from fledging was 2.2 years (Stewart 1952, Henny 1969). This difference has been attributed to the generally more favourable climate, especially in winter, at lower latitudes. However, other factors could also be important. Most northerly populations are dependent upon voles which fluctuate in abundance and, as will be shown below, mortality of the owls is high when vole numbers decline. It is possible that small mammal populations fluctuate less at the lower latitudes, thus providing a somewhat more stable food supply and hence an increased survival rate.

14.3 Mortality rate in relation to food supply and weather

Age specific mortality rates, calculated from ringing recoveries, tell us how mortality changes throughout the life cycle, averaged over many generations, but this masks the variations that might occur from year-to-year, caused perhaps by differences in food supply and weather conditions. This is where the second method of estimating mortality rates, by the disappearance of known individuals from intensively studied populations, comes into its own.

In the Scottish area, using the disappearance rates of ringed birds, I was able to test directly for relationships between annual mortality rates of breeding adults and variations in field vole abundance and weather conditions. As most deaths occurred overwinter, it made no sense to express mortality according to calendar years, so the calculations were made from May one year to May the next, coinciding with the start of breeding and when the birds were trapped each year. Field vole abundance was assessed in April (Chapter 5). Relationships were tested by performing a stepwise multiple regression analysis incorporating overwinter mortality rates, field vole abundance and various weather variables from December to March.

Mortality rates were found to be strongly correlated with vole abundance and showed a distinctly cyclic pattern coinciding exactly with the 3-year vole cycle (Figs. 14.4, 14.5). Survival of the owls was highest at the peak of the vole cycle, was lower in the year of decline, and lower still the following year. This gradation can be explained by comparing it with the known overwinter changes in vole numbers over the course

Scan

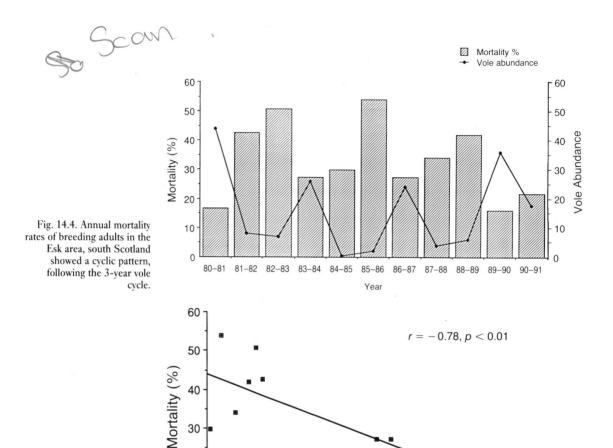

Fig. 14.4. Annual mortality rates of breeding adults in the Esk area, south Scotland showed a cyclic pattern, following the 3-year vole cycle.

Fig. 14.5. Correlation between annual mortality rates of breeding adults and vole abundance in the Esk study area, south Scotland. When voles were scarce mortality rates increased.

of the cycle. In peak years, vole abundance was high throughout the winter, in decline years it was high at the start of winter but declined rapidly towards spring and, in the third year of the cycle, vole numbers were low throughout the whole winter (Chapter 5).

The mortality of first year birds could not be examined in this way but was estimated by calculating the number found dead each year (July to May) as a percentage of the total fledged. Again, mortality was correlated with field vole abundance ($r = -0.78$, $p < 0.01$).

In the multiple regression analysis, none of the winter weather variables was significantly correlated with annual variations in mortality rate. This included the number of days with snow lying, the number of wet days, and average temperatures. Field vole abundance varied independently of winter weather so these results rather surprisingly seem to suggest that year-to-year variations in winter weather were not associated with variations in mortality rates. This is discussed more fully below. In Germany, Schonfeld, Girbig and Strum (1977) also found that mortality rates were related to cyclic changes in common vole abundance and not to weather conditions. Vole abundance again varied independently of winter weather.

14.4 Causes of death

Evidence for the causes of death is mostly circumstantial as so few of the birds that die are ever found. Ringing recoveries suggest a number of possibilities, but, as these are often based on the subjective judgements of members of the public who find the dead birds, they must be treated with caution. Also, what might seem to be the immediate cause of death may not be the ultimate cause. Possibilities reported in recovery reports include: starvation, collision with vehicles, poisoning, shooting and a variety of others including drowning, electrocution, predation, being trapped inside buildings and striking overhead wires. Shooting seems to be a significant problem in some Mediterranean areas (Fajardo 1990). Poisoning from pesticides has been important historically and continues to be so in some regions (Chapter 16).

Starvation

Much of the available evidence suggests that starvation is the main cause of death. For example, of the 66 adults found freshly dead in the Scottish study between 1980 and 1986 (see above), 43 were severely emaciated with no signs of injury or poisoning. Analysis for pesticides was not done but these were rarely used in this largely pastoral area and starvation, with no other complications, seems the obvious candidate. The strong correlation in the Scottish study between vole abundance and mortality rates, which was independent of variations in weather factors, suggests that absolute food shortage was the most important cause of death. However, changes in vole abundance explained only 60% of the annual variations in mortality rates, leaving 40% unaccounted for. The strongly seasonal pattern of mortality, with most occurring in winter (Fig. 14.2), suggests that weather was involved in some way. This seems, at first consideration, to be incompatible with the finding that annual variations in mortality rates were not correlated with annual variations in winter weather. However, it is important to remember that the difference between the weather conditions in the study area in summer and those in winter far exceeded the variations between winters (Chapter 1). In winter, average temperatures were lower, the number of wet days and total rainfall was higher and so also were average windspeeds. Winter also brought sleet and snowfall and days when deep snow covered the ground. As a result, the energy expended by the birds simply to maintain themselves was higher in winter and periods when weather conditions would have inhibited or prevented hunting occurred more frequently (Chapter 6). At the same time these conditions would have lowered the activity of the small mammal prey reducing their availability to the owls. Thus, mortality through starvation could be greater in winter than in summer simply by all these factors working in combination. Yearly variations in vole abundance were superimposed upon this seasonal weather pattern, causing yearly variations in the extent of the winter mortality.

The failure to demonstrate a correlation between overwinter mortality rates and variations in weather between winters does not necessarily mean that it did not exist. More likely, it was the method of expressing or quantifying weather that was inappropriate. Simple overwinter totals of days of snow cover, rainfall or average temperatures may not have adequately measured what was important for the owls. The duration of periods of exceptionally adverse weather conditions is probably of greater significance than these annual totals or averages and the cumulative effect of weather variables, acting independently or in combination, would have to be quantified to assess their impact on the owls' survival. Extremely low temperatures may have less effect over 2 days than over 5 and less on their own than in combination with prolonged strong winds, sleet or snow. The significance of snow cover probably depends on its precise nature as well as duration, how deep it is, whether or not it has a frozen crust and so on. Barn owls can catch prey such as voles through shallow snow (Taylor, personal observation) but deep snow cuts off most of their food supply. Without very detailed information on the owl's foraging behaviour and prey capture rates under all possible conditions it is not possible to decide how these weather variables should be analysed.

Prolonged snow, especially when it covers the ground uniformly and at depths greater than about 7 cm and when accompanied by exceptionally low temperatures, has often been identified with increased owl mortality (Chapter 6, Honer 1963, Güttinger 1965, Kaus 1977, de Jong 1983, Marti and Wagner 1985). The most carefully researched study of this was by Marti and Wagner (1985) in Utah over the winter of 1981 to 1982. Between 1 January and 28 February 1982, 74 severely emaciated barn owls were found dead. Most perished during a period of 2 weeks when there was snow cover 20 to 25 cm deep and mean daily temperatures of −9.7 °C. The overall mortality rate during this period was unknown but the breeding population fell by 40% from 1981 to 1982. It is perhaps more remarkable that so many birds survived these conditions rather than the numbers that died. Events such as this have been reported for the winters of 1962/63 and 1978/79 in Europe (Güttinger 1965, Taylor 1992) but they are exceptional.

Mortality rates can actually be considerably higher than average when there is little overwinter snow and mild conditions but low vole populations. In the Scottish area, the winter of 1988 to 1989 was exceptionally mild with only 7 days of thin snow cover, but the vole populations were very low and the breeding adults suffered a 42% mortality rate compared with the long-term average of 33% between 1980 and 1991. Deaths associated with snow may be conspicuous because they are concentrated into short periods and their importance may therefore be exaggerated somewhat. Only 16 of the 43 cases of starved adults found during intensive searches in my study area could clearly be associated with snow and low temperatures. Eleven were found after periods of prolonged strong winds and heavy rainfall, and 18 could not easily be associated with any particular set of conditions.

A strong correlation was found between altitude and the annual mortality rates of breeding-age adults in the Scottish area. Below 50 m the average was about 20%, rising to around 60% at 250 m. All weather variables, but especially temperature, changed considerably over this altitude range. Increasing mortality rates are probably responsible for the altitudinal limits to the barn owl's distribution, which in Britain is about 250 m to 300 m. Further south, in Spain, this is extended to about 1300 m (Alegre et al. 1989) and in Franconia, Germany to about 500 m (Kaus 1977). Interestingly, in all areas, this limit coincides with about 40 days of snow cover throughout the winter.

Road deaths

Collision with vehicles is the most frequently reported circumstance of mortality in the national ringing recoveries from most European countries. This might suggest that this is now the most important cause of death but these recoveries are strongly biased against birds that die in less obvious situations. There are also few reliable estimates of the total numbers killed by vehicles or of their significance as a proportion of all deaths. An average of 2.6 dead barn owls was found per year along a 36.9 km sample stretch of the Geneva–Lucerne highway sampled over 17 years (Borquin 1983) but as no details were given of the density of the surrounding barn owl population it is impossible to assess the significance of this figure. In Westphalia, Illner (1992) found an average of 0.7 dead barn owls per 100 km of road per year over 12 years which was estimated to be about 10 to 15% of all deaths in the population.

Collision may be the immediate cause of death for many birds but several pieces of evidence suggest that it is often not the ultimate cause. Firstly, most road deaths occur in winter (Borquin 1983, Newton *et al.* 1991) even though the volume of road traffic and probably also average speeds is much greater in summer than in winter, especially in rural areas. Illner (1992) showed that vehicle speed was an important factor. Against this however is the possibility that birds may be dazzled and

(*a*)

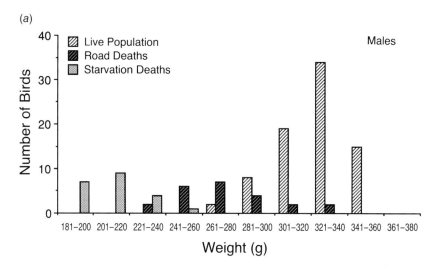

(*b*)

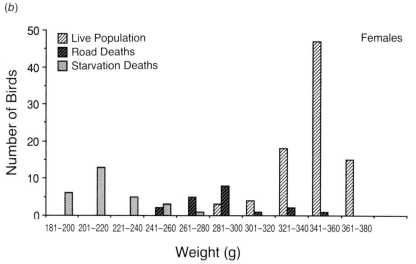

Fig. 14.6. Comparisons of the fresh weights of 'road-killed' barn owls with those of healthy live birds and those that died naturally of starvation. Data for adult birds only, Esk study area, south Scotland for the months of December to February, 1980 to 1992. Most, but not all, of the 'road-killed' birds were in poor condition.

confused by headlights at night, so that the longer winter nights might take a higher toll. Some road victims are within the weight range of apparently normal, healthy birds but most are in poor condition. This has been demonstrated in the Scottish study area (Fig. 14.6) and in Utah (Marti and Wagner 1985) but, of course, might not be the case in all areas. Such light-weight individuals may have been less able to avoid vehicles as they moved across roads from one foraging patch to another or they may have been forced, through lack of alternatives, to feed in situations where they were more vulnerable, such as along roadside verges. It is impossible to say how many of them would have recovered body condition and survived had they not been struck and how many would have died anyway. However, food shortage, whether temporary or prolonged, is implicated and the ultimate cause may have more to do with habitat quality, natural prey abundance and weather than to road traffic.

Several of the birds that we radio-tracked in Scotland had long stretches of wide grass verges along quiet minor roads immediately adjacent to their nest sites. Usually only a strip of about a metre or so wide along these road edges was cut short and the rest had long vegetation and high densities of voles and shrews, yet we never saw the birds hunt these areas. All of the birds we observed were in good body condition.

14.5 Summary

Barn owls living in temperate areas show a distinctly seasonal pattern of deaths with newly fledged birds experiencing high mortality rates in their first autumn and both first years and adults having high overwinter mortalities. Few deaths occur in the summer months. Mortality in these areas is strongly age-related with first-year rates of 65–75% falling to 50–60% and 30–40% in the second and third years. More southerly barn owls have significantly lower mortality rates. In the Scottish study area, annual mortality rates of breeding adults varied from 16% to 55% and were closely correlated with the cyclic variation in field vole abundance. No simple relationship could be found between these annual variations in mortality rates and winter weather variables in this study, but exceptionally prolonged periods of deep snow cover and extremely low temperatures have been associated with increased mortality in many studies. Mortality rate increases with altitude. Starvation, through reduced prey abundance and the effects of weather on prey availability and the birds' energy requirements is suggested as the main cause of death. The significance of road deaths is discussed.

15 *Population size and regulation*

In the preceding chapters we have seen that many barn owl populations, including most of those studied in Europe, show considerable annual variations in laying dates, clutch sizes, fledging success, mortality rates and, in some cases, also in dispersal behaviour. These have all been related mainly to variations in food supply, although extremely severe winter conditions in northern areas have also been shown to increase mortality rates on occasion. Some populations, such as those studied in Utah and New Jersey, have exhibited greater stability in these parameters.

This chapter is concerned with the number of owls: how this varies from year to year, what causes these variations and what, ultimately, is responsible for determining numbers in any particular area. One of the main objectives in my own study was to quantify all aspects of the owls' population dynamics so that the mechanisms responsible for determining numbers could be understood. Unfortunately, no other published studies have been complete enough to allow comparisons with the Scottish study, although the current researches of Bruce Colvin and Paul Hegdal in New Jersey and of Carl Marti in Utah are equivalent in every respect and when published will certainly allow very interesting comparisons. Most studies have one extremely important, but understandable, deficiency: they have not quantified annual variations in the mortality rate of breeding adults. To achieve this the birds have to be trapped at their nest sites so that they can be ringed and identified and it is important to capture them early in the breeding season otherwise the figures obtained will be biased towards successful breeders. This has to be done with great care to avoid the possibility of nest desertions; in my area I was lucky as the birds were not disturbed by my visits (Taylor 1991a) and this seems to be the case in most of Europe, but might not always be so. It is important to check for this in each new population studied. Also, the disappearance rates of ringed adults can only be used to estimate mortality rates when the birds are highly faithful to their nest sites and this is not always the case.

Another constraint in making comparisons among areas is that few studies have located all nests; usually only those in conspicuous sites have been counted, so that absolute densities were unknown. There is also a strong geographic bias, we know little about population sizes in tropical areas which include most of the barn owl's subspecies.

15.1 Nature of population changes

Population changes can be both short term, occurring within the space of a few years, and long term. An examination of short-term changes and their causes can very often lead to an understanding of the mechanisms that are responsible for controlling numbers in the long term and so is a valuable starting point. Many populations have shown long-term declines in relation to land-use changes (Chapter 16) so results can also have a direct conservation relevance. It may become desirable to monitor numbers to detect man-induced changes as part of a conservation strategy and the methods used in this must be based upon a knowledge of how numbers vary in the short term.

In south Scotland, I assessed the number of breeding pairs in my main study area (Esk area) for 14 consecutive years, from 1979 to 1992, and in the two subsidiary areas (Dee and Nith areas, Fig. 1.2, p. 4) for 7 years from 1980 to 1986. The idea of having these additional areas was simply to check whether the changes found in the Esk area were representative of wider regions. Thorough searches were conducted each year in an attempt to locate all breeding pairs in these areas. It was important also to monitor prey numbers each year. This was achieved by trapping in the Esk area but in the Dee and Nith areas, because of time constraints, it was only possible to score field vole abundance as peak, decline and increase phases of the cycle, assessed from the density of occupied runways. The vole cycles were synchronised over all three areas which in some ways simplifies comparisons but it might have been considerably more instructive had they not been synchronised, as this would have provided a valuable natural experiment.

Throughout the 14-year study, the number of breeding pairs in the Esk area varied from a maximum of 55 pairs in 1981 to a low of 15 pairs in 1986. The changes followed an almost exact cyclic pattern with peak owl abundances at 3-year intervals, in 1981, 1984, 1987 and 1990. In each cycle the decline in numbers occurred over 2 successive years so that lowest levels were reached the year after the initial year of decline and the increase phase took place over only 1 year (Fig. 15.1). Changes were synchronised in all habitats: fluctuations in conifer plantations, sheepwalk and pastoral enclosed farmland were correlated. Changes in breeding numbers in the Dee and Nith areas were also strongly correlated with those in the Esk area (Spearman rank correlation: $R_s = 0.99$, $p < 0.01$, and $R_s = 0.85$, $p < 0.01$, respectively) so it is likely that barn owl population fluctuations occurred in synchrony over a wide area of south Scotland.

Elsewhere in Europe there have been several long-term studies of barn owl breeding numbers. In most of these, counts were mainly of

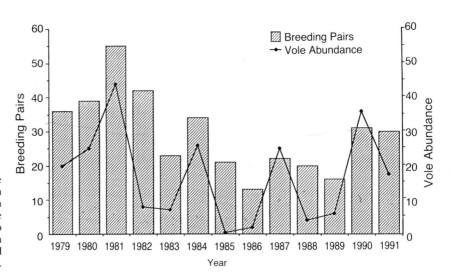

Fig. 15.1. Changes in the number of breeding pairs in the Esk study area, south Scotland in relation to the field vole cycle, from 1979 to 1991. A strongly cyclic pattern is evident in the owl population.

pairs nesting in conspicuous sites such as church towers and attempts were not made to locate all pairs in the study areas. There is a possibility that the figures obtained did not give a totally accurate picture of overall population changes as such sites may have been easier for the owls to find and probably were also more secure from disturbance and predation, so may have been occupied more consistently than other sites. Nevertheless, they probably give a good enough idea of the general trends and minimum levels of population fluctuation in these areas.

The largest continuous set of data, now spanning 18 years, has been obtained from a small study area in Germany in which all pairs were located each year (Illner 1988). Numbers fluctuated against a background of gradual long-term decline, with about a 2- to 3-fold amplitude between peaks and troughs and, although there was a tendency towards a cyclic pattern with a 3- to 4-year periodicity, this was not as convincingly regular as that found in the Scottish population (Table 15.1). Two other German studies also showed considerable year-to-year changes, with 4 to 5-fold fluctuations in numbers (Table 15.1, Kaus 1977, Schonfeld and Girbig 1975), and similar results were obtained from two studies in France (Baudvin 1986, Muller 1989). Although there were some indications of regularity in the fluctuations recorded in all of these studies, they cannot clearly and conclusively be described as cyclic on the basis of present information. Longer runs of data are needed that can be subjected to objective statistical analysis.

Two North American populations that have been studied over many years have shown greater stability from year to year than the European populations. In New Jersey, the study area supported an average of around 50 pairs between 1980 and 1990, with maximum and minimum numbers of 63 and 44 pairs. This represents only about a 1.5-fold variation in the amplitude of fluctuations compared with the more normal 3- to 5-fold variations found in northern and central Europe. There was also no indication at all of cyclicity in the New Jersey population (Colvin 1984, Colvin and Hegdal 1986, 1987, 1988, 1989, 1990). Breeding numbers in a Utah study population also varied little from year-to-year generally, again with no evidence of cyclicity. The main deviations away from average values were associated with periodically exceptionally severe winters (Marti and Wagner 1985).

15.2 Causes of population changes

The most striking characteristic of the year-to-year population fluctuations in the Scottish study of owls was their distinctly cyclic nature. Owl numbers followed the vole cycle closely so that spring owl and field

Table 15.1. *Annual variation in the number of breeding pairs in five continental European studies*

Year	Number of breeding pairs in study				
	1	2	3	4	5
1966	22				
1967	36				
1968	7	50			
1969	12	15			
1970	8	32			
1971	14	54			
1972	33	35	64		
1973	3	18	4		
1974	17	27	77	18	
1975			57	26	
1976			33	10	
1977			53	22	0.73
1978			40	18	0.64
1979			19	3	0.35
1980				11	0.52
1981				17	0.81
1982				8	0.42
1983				10	0.78
1984				12	0.62
1985				6	0.33
1986				9	0.55
1987				7	0.33
1988				8	0.67

The values given in study 5 were an 'index of nesting', i.e. the proportion of nest sites visited that held breeding pairs.
Studies taken from: 1, Kaus 1977; 2, Schonfeld and Girbig 1975; 3, Baudvin 1986; 4, Illner 1988; 5, Muller 1989.

vole abundances were clearly correlated (Figs. 15.1, 15.2). Just as annual changes in vole abundance were not correlated with overwinter weather conditions (Chapter 5), so also the population changes in owls were not significantly associated with any weather variables. The population change in owls, therefore, seems to have been predominantly in response to changes in food supply.

The mechanism for this response could potentially have involved

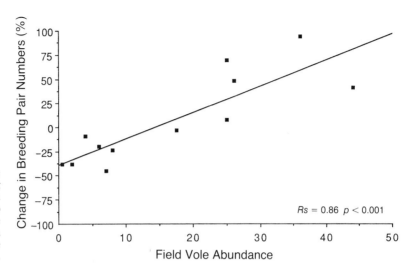

Fig. 15.2. Percentage change in the number of breeding pairs in the Esk study area between years in relation to spring vole abundances. The number of owls showed a rapid tracking of the vole population.

variations in the production of offspring, immigration rate, mortality rate or changes in the proportions of birds available each spring that were recruited to the breeding population. The nature of the population changes precludes the possibility that they were caused primarily by fluctuations in the production of young: years of decline were always preceded by years of maximum breeding output and years of increase followed years of low population productivity. Annual changes in breeding numbers were not simply correlated with the number of young fledged the previous year ($r = -0.35$, $p > 0.10$). Movements were not a significant feature of barn owl populations in south Scotland and there was no evidence of variations in the dispersal distances of first year birds in response to the vole cycle (Chapter 13). Also, among the breeding adults trapped each year there was no significant variation in the proportion that had been ringed as fledglings in the study area and those that were unringed and that might therefore have been immigrants.

In Chapter 14 it was shown that the mortality rates of both breeding adults and of first year birds were significantly correlated with annual changes in field vole abundances. Variations in the mortality rate of adults were substantial, from about 20% to 56%, between years. It is therefore perhaps not surprising to discover that the yearly variations in breeding numbers were also strongly correlated with the overwinter mortality of breeding adults (Fig. 15.3). This factor alone predicted about 60% of the total variation in breeding numbers. Population stability was achieved with a breeding adult annual mortality rate of 35%; above this the population declined and below it, the population tended to increase.

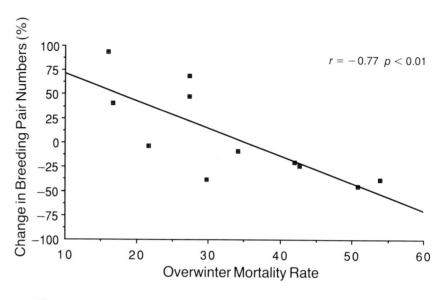

Fig. 15.3. Percentage change in the number of breeding pairs between years in the Esk study area, south Scotland in relation to the overwinter mortality of the breeding birds.

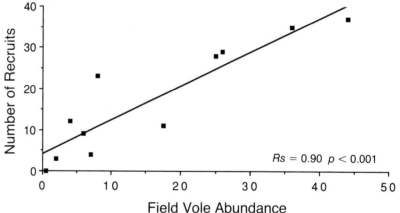

Fig. 15.4. The number of birds recruited to the Esk population each year in relation to the spring field vole abundance. When voles were abundant, more birds were added to the breeding population.

The total number of individuals recruited to the breeding population each year was also closely correlated with field vole abundance: when voles were numerous, large numbers were recruited and when voles were scarce, few were recruited (Fig. 15.4). The majority (63.4%, $n = 186$) of the new recruits recorded were first year birds breeding for the first time but there was also a substantial number of former breeding adults that had not bred the previous year (32.8%), usually because they had lost their mates and failed to obtain replacements, and a very small number of individuals that were trapped for the first time as breeding adults in their second year of life and that had not been recorded as breeders in their first year (3.8%).

Table 15.2. *The number of first year birds recruited to the breeding population in relation to the time of year that they fledged from the nest. Those that fledged before median fledging date for their year had a higher chance of being recruited*

	Fledging date of breeding first year birds	
	Before median date	After median date
Number of males	29	8
Number of females	31	15

The young birds' chances of being recruited to the breeding population depended very much on when they fledged from the nest in any particular year. For each year of the study, I recorded whether successful recruits that had been ringed as nestlings in the study area had fledged before or after the median fledging date of all ringed nestlings in their particular year. Both males and females that fledged before the median date (i.e. the date at which half the population had fledged) had a much higher probability of being recruited to breed than those that fledged after the median date (Table 15.2). Presumably, those that fledged early had more time to improve their hunting skills before winter and survived better. This demonstrates the importance to barn owls of laying as early as possible.

The annual variation in the recruitment rate of first year birds probably resulted mainly from differential survival rates during their first winter (Chapter 14). However, in addition to this, there were also variations in the proportion of those that survived the winter but that then either joined the breeding population or spent the summer as non-breeding single birds. Non-breeding individuals were exceedingly difficult to identify as such. Any bird found roosting during the spring or summer up to 3 km from an occupied nest could have been either the female or more likely the male from that nest. Thus the only way to be sure that an apparently single bird was a non-breeder was to capture and identify it, at the same time as locating all adjacent nests and indentifying the breeding birds. It was usually considerably more difficult to catch single roosting birds than birds at the nest and a useful short-cut to identify them was to dye-mark breeding adults so that any non-breeders could be recognised by the absence of marks without having to capture them.

Identifying non-breeders with certainty was a time consuming task and also, because the variety of sites used by them for roosting was highly diverse and often not at all obvious, it was never possible to estimate their total numbers. Nevertheless, for 9 of the study years I searched a standard set of potential roosting places (which also included potential nesting places) and by a combination of trapping and the use of dye-marking established the number of non-breeders using them. This at least gave an index of their abundance each year which could then be related to vole abundance. In vole peak years hardly any non-breeders were found whereas in low vole years substantial numbers were found (Fig. 15.5).

It is therefore possible to identify three population processes that were responsible for year-to-year changes in breeding numbers in the Scottish population: (1) the survival rate of breeding adults, (2) the survival rate of first year birds and (3) the rate of recruitment of birds of all ages to the breeding population. All three varied in response to the cyclic changes in vole abundance.

The winter immediately preceding the start of the study in south Scotland (1978/79) was one of the most severe ever recorded in the area. At the highest point of the study area, around Eskdalemuir, 62 days were recorded with snow lying on the ground and at the lowest point, at Carlisle, there were 24 days with snow cover, twice the long-term average values. These snow days occurred mainly in January and February (65%) and there were prolonged periods of continuous deep snow cover during these months. Mean daily temperatures at Eskdalemuir were $-2.1\,°C$ for January and $-0.5\,°C$ for February, $3.5\,°C$ and $2.0\,°C$ lower than the long-term averages.

With information from a number of local ornithologists throughout the county of Dumfriesshire, immediately around the study areas, I was able to compare the occupancy in 1979 of 80 nest sites in which the owls bred in 1978. Breeding was recorded at 48 of these sites in 1979,

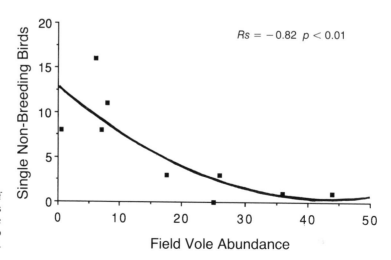

Fig. 15.5. The number of non-breeding single birds discovered each year in the Esk study area in relation to field vole abundance.

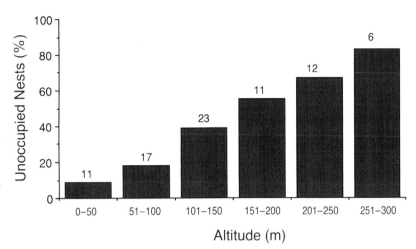

Fig. 15.6. The loss of breeding pairs at 80 nest sites around the Esk and Nith study areas following the exceptionally severe winter of 1978/79. Those at lowest altitudes suffered the lowest losses. The number of nests is shown above each column.

apparently single birds at 17 and no birds or signs such as fresh pellets at 15 of them. Losses were strongly correlated with altitude (Fig. 15.6). Considering only sites between 0 and 150 m above sea level, the range within which most British barn owls nest (Bunn *et al.* 1982), breeding occurred at 51 in 1978 and at 38 in 1979, a decrease of only 25.5%. By 1981 breeding pairs were re-established at 49 (96%) of these sites. Thus at these altitudes the effect of this exceptionally cold and snowy winter on the number of pairs was relatively minor and short-lived.

Nigel Charles of the Institute of Terrestrial Ecology assessed field vole abundance within the Esk study area during 1978 and 1979 (see Chapter 5) and found that populations were relatively high throughout; the prolonged snow and low temperatures were not followed by a decline in vole numbers in 1979. The changes in the owl population can probably be attributed directly to the weather conditions reducing prey availability at the same time as increasing the birds' energy needs, resulting in reduced survival. It seems, therefore, that in addition to the more usual cyclic fluctuations in numbers brought about by variations in food supply there were also periodic reductions with subsequent recovery caused by exceptionally severe winter weather. Such events are relatively rare but completely unpredictable. As such they present a very serious problem to any monitoring programme that is not based on continuous annual assessments.

The 1978/79 winter seems to have affected barn owl populations throughout much of Europe; substantial declines were noted in France (Baudvin 1986, Muller 1989), Switzerland (Juillard and Beuret 1983) and Germany (Illner 1988), but in all cases recovery occurred rapidly, within 1 or 2 years.

Fluctuations in barn owl numbers in continental Europe have long been associated with changes in the abundance of common voles. Honer (1963) compiled a great deal of evidence that linked what he termed 'mortality bursts' (or *Sterbejahr*, Sauter 1956) in the owl population with peaks and declines in vole numbers and inferred that owl numbers fluctuated in response to vole abundance. Unfortunately, there have been few studies in which owl breeding numbers and vole numbers were measured directly and simultaneously and this has given rise to considerable confusion in the literature over the relative importance of vole abundance in comparison with severe winters as causes of fluctuating owl numbers. In Frankonia (part of Bavaria), peaks in owl abundance were correlated with peaks in common vole abundance between 1966 and 1974 and years of decline were also correlated (Kaus 1977). In Hubertus Illner's study area, again in Germany, between 1974 and 1991, years of high vole abundance were also years of high barn owl breeding numbers (Illner 1988 and personal communication). Schonfeld and Girbig 1975 and Schonfeld *et al.* 1977 assessed owl numbers, vole abundance and winter weather conditions in their study area in the former East Germany between 1968 and 1973 and found a correlation between owl and vole abundance but no clear association between either of these and winter temperatures or the number of days with snow covering the ground. By contrast, in France, Muller (1989) found that years of lower than average owl numbers between 1977 and 1988 followed winters with below average temperatures and above average snow cover (>40 days). Vole abundance was not estimated so no direct comparisons could be made with weather conditions. Muller concluded that winter weather was the most important factor controlling the abundance of the owls. However, the situation might not be quite so clear cut. Owl breeding numbers actually increased by 65% between 1985 and 1986 when the intervening winter had by far the greatest duration of snow cover and lower than average temperatures. Also, although vole abundance was not assessed, data were presented on the breeding success of the owls each year (Muller 1990) from which it is possible to infer something about vole abundance. Every study that has been conducted in Europe has shown that the productivity of first broods and the incidence of second broods are closely correlated with vole abundance (Chapter 12). In Muller's study, 2 of the 4 years in which owl numbers declined following winters with worse than average weather conditions had been preceded by years in which the productivity of first broods and the incidence of second broods had fallen sharply. Breeding numbers had also declined, all of which suggests that the vole population had declined. It seems likely that the owl population began these winters with already low food supplies and falling numbers. Probably, the fluctuations in owl numbers over the study can only be adequately explained by considering both vole abundance and weather conditions.

The two North American studies showed no evidence of cyclic changes in barn owl abundance and numbers fluctuated rather little from year-to-year compared with European populations. Vole populations in these areas were not cyclic; so it seems that where vole abundance is more stable, barn owl numbers are also more stable. The main changes in these owl populations were periodic declines which could be associated with extremes in weather conditions. In New Jersey the only large-scale decline involved a fall from 63 pairs in 1986 to 44 pairs in 1987. The spring and summer of 1986 experienced extreme drought conditions in the study area, apparently the worst in living memory, which severely reduced the growth of vegetation and, consequently, the population of voles. As a result the breeding success of the owls was considerably reduced. Vole populations overwinter were reduced and survival of the young birds was low so that few were recruited to the breeding populations in 1987 (Colvin and Hegdal, 1986, 1987). In Utah, winter conditions seem to exert the greatest influence on the owl population, with numbers declining after exceptionally severe winters, such as those of 1981/82 and 1985/86. In this area winters can be considerably more severe than any experienced within the owls' range in Europe, with up to 100 days of snow cover and temperatures far below freezing (Marti and Wagner 1985, Carl Marti, personal communication).

In other parts of the world there is much less quantitative information on barn owl population changes in relation to food supply or weather. Numbers declined in a study area in Mali following a fall in the abundance of rodents (Wilson *et al.* 1986) and from Australia there are many anecdotal accounts of barn owls declining after the collapse of house mouse 'plagues' (Fleay 1968, Schodde and Mason 1980, Hollands 1991).

A most striking feature of all barn owl populations that have been studied in Europe and North America is their ability to recover rapidly within 1 or 2 years following even very large-scale reductions. It is likely that changes in survival rates are important in this but the recruitment of a previously non-breeding component of the population is probably also important. Non-breeding birds have been recorded in many studies (e.g. this study, Marti and Wagner 1985, Muller 1989) but their numbers have never been adequately quantified. Most conservation programmes concentrate almost exclusively on the breeding population when they should also be considering the fate of this important non-breeding segment. The survival rate of these birds and the factors affecting it are largely unknown.

Some of the recoveries in population numbers recorded in continental

Europe seem however to have been too rapid to be explained simply by changes in survival and the recruitment of previous non-breeders and it is possible that immigration has also been important for these cases. If this is indeed so, there must be significant conservation implications as attempts to conserve the birds in one area may not be successful unless similar conservation measures are carried out more widely. It is important to learn what the connections are between the birds in different areas and this argues for more intensive programmes of ringing and capturing breeding adults.

15.3 Limitation of population density

The preceding section considered how barn owl numbers change from year-to-year and identified some of the factors responsible for those changes, but it did not explain why numbers should fluctuate around a particular average level rather than around some other level. Why, for example, did the Esk study area in south Scotland not support, on average, twice the density or half the density of owls that it did? Newton (1979) identified food supply and nest sites as the two factors most likely to be responsible for the long-term regulation or limitation of raptor densities. Populations may be regulated around particular densities by food supply such that if their density is increased above the optimal level there may be competition among individuals resulting in reduced recruitment rates, higher mortality rates or lower reproduction rates. Such regulation is termed density dependent since its intensity depends upon the density of the birds. Often among raptors the breeding birds are strongly territorial and individuals or pairs compete indirectly for food by competing for space. Territory size may vary from place to place according to prey abundance (e.g. in Sparrowhawks, Newton 1986) or from year-to-year in relation to changes in prey abundance (e.g. in kestrels, Village 1990) and, thus, the density of pairs may be adjusted to the level of food supply.

The information we have for barn owls indicates that they do not normally defend large areas for foraging. During the summer they defend their nest sites strongly and probably also their roost sites but feeding areas overlap extensively. Therefore it seems highly unlikely that the density of pairs in any area is limited by social competition for feeding areas. This leaves two other main possibilities: that the birds compete directly for resources such as food so that as their density increases their breeding success is lowered, their mortality increased or their emigration rate increased thereby limiting the potential for further rises in density, or that their numbers are limited by the availability of nest sites in suitable foraging habitat.

Barn owls respond very rapidly to changes in their food supply with the result that their numbers are almost always closely correlated to those of their prey. This makes it extremely difficult to test for the existence of density-dependent effects within their populations and to date no one has succeeded in achieving this. However, in my study area there were many pairs in the plantation and sheepwalk habitats that were a long way from other pairs and where the possibility of any significant direct competition for food seemed unlikely. In these areas, nest sites were few and far between and it was quite obvious that in some years, such as 1981, when the owl population was at its highest there were simply few unused nest sites. In natural habitats, nest sites with the exact specifications needed by barn owls are never likely to be abundant and to be used by the owls they also have to be in areas of suitable foraging habitat. There is likely to be intense competition for cavities from a whole range of mammals and other birds which would further restrict the possibilities for the owls. Nest sites that are suitable one year may not be suitable the following year if prey abundance falls below a certain level. Therefore it is possible that the availability of nest sites in conjunction with prey abundance serves to limit the maximum density of owls in any area. This limitation may only come into effect periodically following prolonged periods of good food supplies.

15.4 Summary

Throughout most of Europe, barn owl breeding populations fluctuate greatly from year to year, with up to a 3- to 5-fold amplitude. These fluctuations tend to be correlated with variations in food supply and extremely severe winter weather conditions. In the Scottish study, the fluctuations were strongly cyclic, following the 3-year field vole cycle. Populations that have been studied in northern areas of the USA, where prey abundance varied less from year-to-year, showed no evidence of cyclic changes and fluctuations were generally of lower amplitude than those found in Europe. In the Scottish study annual changes in breeding numbers were the result of variations in overwinter mortality, through starvation, among the breeding birds and first year birds and were related to vole numbers. Variations in recruitment rate were also important. Barn owl populations usually have a non-breeding component that is important in the dynamics of the population. It seems likely that in natural situations, the availability of suitable nest cavities in suitable foraging habitat is responsible ultimately for limiting the density of the owls.

16 *Conservation*

F ew other birds have such a wide distribution as the barn owl and globally it is still an extremely abundant species. Only in some parts of Europe and North America is there a genuine conservation problem, although some of the subspecies inhabiting smaller islands might also be experiencing difficulties. Unfortunately, for these we have no published information. The extent of population decline, its exact causes and possible remedies vary considerably from place to place and it is difficult to provide an exhaustive treatment covering all regions within the scope of this book. My intention therefore is to discuss the main principles involved and to illustrate some of the details.

16.1 Evidence for population decline

Barn owls are unobtrusive, inconspicuous birds and are difficult to track down, especially when they are at low densities. Some nests may be found easily, but to discover all pairs in an area is a very difficult and time-consuming task. Added to this is the dynamic nature of barn owl populations, involving large short-term changes in breeding numbers (Chapter 15). Together, these difficulties mean that we have almost no quantitative evidence for long-term population declines that stands up to critical scrutiny.

The most reliable evidence we have is at a somewhat simpler level and more often than not anecdotal. Throughout Europe and North America the first-hand experiences of amateur ornithologists, farmers and a few professional biologists all point inarguably to prolonged and often large-scale declines starting in many areas in the 1930s or 1940s and accelerating during the 1960s and 1970s. For Britain, information of this kind was gathered by Prestt (1965) and Bunn *et al.* (1982) and many regional ornithological and natural history societies and individual researchers over Europe have published similar evidence (e.g. Güttinger 1965, Krägenow 1970, Straeten and Asselberg 1973, Braaksma and de Bruijn 1976, de Wavrin 1977, Ziesemer 1980, de Jong 1983, Illner 1988).

In the United States, barn owls seem to have moved into some of the more northerly states only with the clearing of forests and general development of agriculture following European settlement. The irrigation of former dry areas also allowed expansion. Populations in Ohio seem to have peaked in the 1930s and declined thereafter (Colvin 1985) and, generally, most mid-west states have experienced long-term reductions since about that time (Stewart 1980, Lerg 1984, Mumford and Keller 1984). The barn owl has now been placed on the endangered species list or cited as a species of special concern in many states (Arbib 1979, Anonymous 1984, Tate 1986).

We will never know with any degree of certainty exactly how large and extensive the population declines have been since the 1930s in Europe and North America but this is not neccesarily the important point. The main strength of the case for conservation comes from the overwhelming mass of anecdotal testimony, originating mainly from farmers and natural historians, that demonstrates beyond any conceivable doubt that there has been a dramatic decline in numbers in many parts of the barn owls' former range.

16.2 Causes of population decline

The reasoning behind an interest in the possible causes of decline is simple: if we can identify causes we might be able to rectify them and perhaps reverse current population declines. However, it must be remembered that this is not the only and probably not the best approach to arrive at a conservation prescription of the species. A more sound alternative is to conduct detailed long-term ecological studies of existing populations in present day landscapes and from these gain an in-depth understanding of the species' requirements from which management guidelines can be drawn up. Nevertheless, consideration of causes may help to identify some points that would otherwise not be obvious and may also be useful in the political arena of conservation.

The reasons for changes in a species' status are most easily elucidated when temporal and spatial correlations can be examined. In other words, when runs of data are available over the same period both on the numbers of the bird and on changes in the possible causal factors. Alternatively data from many populations over which possible factors have operated but at differing intensities can be studied. Such approaches formed a successful and important component in the development of the case against the use of organochlorine pesticides (Newton 1986). Unfortunately, in the case of the barn owl, the detailed historical information on population numbers that is needed to make good use of this approach is not available. Nevertheless, it is possible to identify what seem to have been one or two turning points for many populations. The first was over the period roughly from the 1930s to the 1940s and the second from the late 1950s to the mid-1960s onwards. These correspond very approximately with the beginnings of modern agricultural methods, particularly the development of farm machinery and consequently the intensification of farming, and then the introduction of organochlorine pesticides. An explanation for the decline in barn owl numbers is probably to be found in the nature and extent of these and subsequent changes to the farmland environment, and it is worth considering in some detail how these may have affected the owls.

Loss of foraging habitat

In the areas under consideration in Europe and North America, barn owls depend upon relatively few prey species, mainly voles, shrews and various species of mice which they catch mostly in grassland habitats (Chapters 3 and 6). Historically, commensal rodents and birds seem also to have been important and continue to be so in some areas. It is evident that the owls would most likely be affected by changes in the abundance, distribution and quality of grasslands within the farmland environment and by changes that influence the abundance of commensal species.

Before the 1930s and 1940s most farms were mixed enterprises with livestock and a variety of crops and had evolved as self-sustaining, integrated systems. Horses provided the power for ploughing, harvesting and transport. With no inorganic fertilizer, manure from livestock was needed to maintain fertility and hay, root crops and cereals were grown to feed the livestock. Crop rotations, often involving cash crops, legumes and fallow grassland were essential in the fertility cycle and in the fight against crop pests. The diversity of crops and the dynamic nature of the farming programme encouraged the development of relatively small field units, separated by hedges, ditches and fencerows. Small woodlands were valuable resources providing fuel, fencing and building materials. Thus, most farmland offered a mosaic of habitats with a high percentage of edge habitat types and, particularly, a high percentage of lightly grazed or ungrazed grassland.

The harvesting and storage of cereals and root crops was relatively inefficient and attracted large populations of 'pests', particularly rats and mice. Sheaves of cut corn were assembled into stacks (or corn-ricks) which were set alongside the barns in stackyards (Fig. 16.1). Threshing was done when corn was needed so the ricks often persisted for most of the winter giving shelter and food supply to extraordinarily high densities of rodents and sparrows (Southwick 1958, Rowe, Taylor and Chudley 1963, 1964). The older farmers I talked to had good memories of the stackyards and recalled seeing barn owls hunting there frequently in winter. These may have been a very important source of food to the owls, perhaps when snow cover reduced the availability of prey in the field habitats or when prey was scarce at the low points of vole cycles. It was shown in the previous chapter that cycles in owl numbers occurred mainly as a result of changes in the levels of overwinter mortality in relation to the vole cycle; when voles declined, mortality increased. The availability of commensal rodents throughout winter may have buffered the owls against this and may have enabled them to maintain much higher population levels regardless of the vole cycle. This may have provided greater resilience to the effects of very severe winters and may also have allowed the birds to maintain themselves in suboptimal habitats such as those at higher altitude.

From the 1940s onwards, there has been a dramatic revolution in agricultural methods with the rapid spread of increasingly sophisticated machinery, the introduction of inorganic fertilizers and organic pesti-

Fig. 16.1. Before the advent of combine-harvesters, traditional stackyards supported high populations of rodents and were probably important winter feeding areas for the owls. (Photograph: Professor Alexander Fenton.)

cides and the advancement of plant and animal breeding programmes.

The development of machinery has had a profound effect on farmland landscapes. To gain the maximum benefit from their use, field units were enlarged, usually with financial assistance from governments, resulting in a widespread loss of edge habitats such as hedgerows, ditches, small woodlands and damp rough grassland corners and hollows (e.g. in Britain, Nature Conservancy Council 1984). This all required investment from farmers and the intensity of these changes has been highest in the better quality, more profitable land, especially where climates are more favourable to crop growth. Mechanisation has made possible widespread drainage, again reducing the extent of damp, rough grasslands. Horses were, of course, largely displaced so that farms no longer needed hayfields and pastures to feed them. One of the most significant developments was that of the combine harvester. This enabled cereals to be harvested, threshed and transported to drying machines and concrete or steel storage silos with minimum loss. Stack-yards disappeared quickly and the storage of cereals on many farms declined. No quantification has been made, but populations of commensal rodents and seed-eating birds around farms must have declined dramatically. Most of the older farmers I have talked to agree with this.

The traditional methods of growing cereals in Britain and some other countries was to plough and to plant in the spring. This resulted in overwinter fields of stubble, with spilled grain, weed seeds, grasses and some cover which supported populations of rodents and seed-eating birds. In most lowland areas, this practice has been progressively replaced by autumn ploughing and sowing; the stubble habitat is no longer available to the rodents or owls.

The introduction of inorganic fertilisers and organic pesticides removed the need for crop rotations to maintain fertility and control pests and resulted in the loss of much fallow grassland. More importantly, since livestock were no longer needed for manure, the opportunity arose for farms to specialise in particular crop types and whole regions began to grow specific crops such as cereals. The original diverse mosaic of habitats which always included grassland was lost from most areas. The management of grasslands has gone through marked changes designed to increase the production of milk and meat. Most grassland is now worked on short rotations and is ploughed in regularly to be replanted with a restricted range of nutritious, digestible grasses, especially rye grasses, *Lolium* spp. This interrupts the natural invasion of other plants and prevents the development of rough, tussocky or hummocky patches and leads to a level, uniform habitat. The use of fertilizer and extensive drainage allows stocking densities to be increased which again results in a more uniform closely cropped sward and this can be further exaggerated by the use of strictly controlled grazing systems such as strip grazing. As a result, most grasslands now offer less cover for shrews, voles and mice. Species such as the field vole in Britain and the meadow vole in North America have probably been severely affected, although common voles in Europe, because of their burrowing habitat, can still make some use of the closely managed grassland.

The progressive replacement of hay by silage for winter fodder has been significant. Hayfields, by being harvested towards the end of the growing season, provide a reasonably stable summer habitat for small mammals and, when they are harvested in late summer, the small mammals must then concentrate into the remaining edge areas, which produces very high densities of prey just at the time the young owls are learning to hunt. Silage fields are usually heavily fertilised, rolled in the spring to encourage tillering, and subject to two and sometimes three harvests per year, creating a hostile environment which supports few small mammals (Chapter 5).

Most studies of small mammals in temperate areas have been concerned with problems of population dynamics and researchers have tended to concentrate on habitats such as woodlands or old fields which offer relatively stable systems. Much less attention has been paid to the dynamic systems of agricultural crops or edge habitats so we have very little thoroughly researched, quantitative information on the effect of these agricultural changes on small mammal populations. Nevertheless, we have a reasonably good understanding of the general ecology and habitat requirements of most of the species concerned and there can be little doubt that their abundance must have declined substantially. Even though we are unable to provide a rigorous quantification, it seems highly probable that falls in barn owl numbers in the more intensively farmed areas have been caused in large part by a reduction in food supply for the owls brought about mainly by the loss of good quality habitat for small mammals.

The loss of nest sites

The loss of nest sites has often been proposed as one of the main factors responsible for declining barn owl numbers in Europe and North America (Krägenow 1970, Braaksma and de Bruijn 1976, Kaus 1977, Bunn *et al.* 1982, Juillard and Beuret 1983). When nesting in buildings, the owls tend to select places that are only infrequently visited by humans (Chapter 10). Traditional farms before the agricultural revolutions of the 1950s and 1960s had many such places, small lofts and sheds that were used to store equipment and barns to store hay and straw (Fig. 16.2). The latter, particularly, seem to have been extremely important. All farms used horses or bullocks, necessitating a large store of hay for winter feed, which was often kept for a year or two to mature. In my areas, hay was mostly stored in stone buildings which had ventilation slits through which the owls could gain access. These lofts were usually left undisturbed during the owls' breeding season and provided good nesting sites. Everywhere, traditional farm buildings, especially barns, are disappearing to be replaced by large multipurpose, open sheds that offer no nesting places. Also in many regions, small farms have gradually become non-viable and have either been abandoned to decay or refurbished as second homes. Throughout continental Europe there has been a growing tendency to wire-off church towers to exclude pigeons but, of course, the owls are also kept out.

Tree nest sites have also been disappearing. Traditional landscapes, evolved before the 20th century, had many large trees, deliberately planted around farmsteads, in parkland and in hedgerows. Many have been lost during field enlargement programmes and generally through

Fig. 16.2. A male barn owl delivers prey to his nest in the loft of an old stone haybarn. Traditional farm buildings are disappearing throughout Europe and North America. (Photograph: Robert T. Smith.)

decay and old age. Once a tree has developed a large enough cavity for barn owls its life span is limited, although some hardwoods, especially oaks, can persist even for centuries with very large cavities. The spread of Dutch elm disease throughout Europe has almost eliminated elms, one of the most frequently used tree species and the ancient practice of pollarding trees which led to the formation of many suitable cavities is now gone.

While nest site availability is almost certainly declining in many areas, there is remarkably little quantitative information on the rate of loss or its impact on barn owl populations. This is understandable as gathering the data necessitates locating all suitable nest sites, not just those actually in use, and is therefore very time consuming. I attempted to do this in my four study areas in south Scotland. Over a period of about 9 months in 1979 and 1980, I searched all buildings and man-made structures that had any potential as nesting places and aided by many helpers examined trees around farms, along roadsides and edges of woods as well as isolated individuals in fields. The criteria described earlier, (Chapter 10), were used to decide whether or not a particular structure was suitable for nesting.

The great majority of farms had no buildings that were even remotely suitable for the owls. In late winter and early spring, when the owls were settling into their nest sites, large numbers of livestock were still in their winter housing and areas that were used for food storage were visited almost every day by the farmers so that almost all parts of the farm were subject to considerable disturbance. However, totally disused and derelict buildings that provided good nest sites were easily identified. Each year over the following decade, all such sites were visited to determine the numbers remaining. Average rates of loss in the south-west study areas (1 to 3) were about 4 to 6% per annum. Thus, 42.9% of sites were lost from Area 1 in 13 years, and 30.3% and 25.0% from Areas 2 and 3 in 7 years (Fig. 16.3). Losses were caused by the complete collapse or demolition of buildings (37.5%), renovation or alteration (53.6%) and windblow of trees (8.9%). Study area 4 lost 60.8% of its 49 potential nest sites in 11 years. Most were roadside elm trees that were killed by Dutch elm disease in the late 1970s and early 1980s and quickly felled for safety reasons (Taylor 1991b).

The effect of these nest site losses on the barn owl population was complex involving some gradual local reductions in the density of breeding pairs, depending upon the original density of available sites. Some areas with high initial numbers of sites suffered no long-term reduction whereas other areas saw the complete loss of nesting pairs. Overall, the

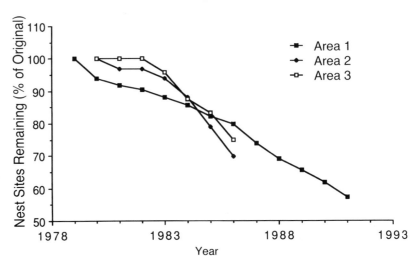

Fig. 16.3. Rate of loss of traditional barn owl nest sites in buildings and trees in the three south Scotland study areas. Nest site loss is now the main threat to these populations.

number of available sites in suitable habitat exceeded the number of owl pairs until 1991 when, had we not started providing large numbers of nest boxes, the owls would have become limited by the number of nest sites available to them. Assuming the study areas to be representative of the whole of south Scotland, this means that a shortage of nest sites is now the main threat to the owl population in this area.

The relative insecurity of tree nest sites was also noted by Colvin (1984) who recorded a 32% permanent loss of such sites ($n = 44$) in only 4 years in New Jersey.

The effects of pesticides

Large-scale declines of many raptors during the 1950s and 1960s have been attributed to the effects of pesticides, particularly to organochlorines which entered the birds' bodies in their food supply and accumulated in their tissues. There is now conclusive evidence from a combination of field data and laboratory experiments that DDE (the metabolite of DDT) causes egg breakage and embryo deaths in many species, resulting in a reduction of breeding performance (Newton 1979, 1986). Breakages are a result of shell thinning caused by the inhibitory effect of DDE on the enzyme carbonic anhydrase, which is necessary for normal shell formation (Peakall 1970). The extent of thinning varied within populations and among different species but generally populations that exhibited an average of more than 16 to 18% shell thinning declined in numbers (Newton 1979). DDT is very stable and hence persists in the environment for many years after application. Therefore, although controls on its use were eventually introduced, recovery was often slow and in some species, such as the sparrowhawk, DDE levels in the birds remained high during the 1980s and continued to reduce their breeding performance (Newton 1986).

Another group of organochlorine pesticides, the cyclodienes, especially the substances aldrin and dieldrin, were also severely damaging to wildlife during the same period and were more important than DDT in the decline of some species. These highly toxic substances killed birds directly, often by damaging their brain tissue, and were probably responsible for many deaths indirectly through the impairment of the birds' hunting efficiency.

The effects of organochlorines were most severe in fish- and bird-eating predators, partly because the chemicals could easily enter their food chain but also because the many steps in the chain provided ample opportunity for concentrations to build up. Over much of their range in Europe and North America, barn owls are associated with grassland habitats where they depend upon voles which do not come into contact with pesticides to any great extent. However, in cereal growing areas, the owls regularly consume commensal rodents such as house mice or brown rats, and various other mice such as the wood mouse, as well as occasional birds, all of which can accumulate pesticides by eating treated cereal seeds or insects (Jeffries and French 1976).

Evidence for the effects of organochlorines on barn owl populations is much more limited than for many other species, partly because of the difficulties involved in assessing the numbers of owls. Nevertheless, the chemicals have been found widely in owl tissues. In New Jersey, all of 18 clutches examined in 1972 and 1973 had residues of DDE and 78% of them showed evidence of dieldrin (Klaas *et al.* 1978). On average, egg shells were 5.5% thinner than pre-1947 shells in the same area. Six (33.3%) clutches had DDE residues greater than 5 p.p.m. and their shells were 14% thinner than pre-1947 samples. Captive barn owls given food with 3.0 p.p.m. of DDE suffered significant egg shell thinning, egg breakage, embryo mortality and reduced production (Mendenhall, Klaas and McLane 1983). Thus, some loss through egg breakage or chick mortality might have been expected in the wild birds, although only one clutch showed evidence of this. DDE residues were recorded in British barn owls but generally at low levels (Cooke, Bell and Haas 1982).

Evidence for aldrin/dieldrin poisoning is much stronger. Of a large sample of barn owls found dead between 1963 and 1989 in Britain, 9% had died of HEOD (the metabolised product of aldrin/dieldrin) poisoning. During the main period of aldrin/dieldrin use up to 1977, the overall value was much higher at 17% and in some cereal growing areas in eastern England where pesticide use was highest, up to 40% of the owls examined may have died in this way. None of the sample of 25 owls collected in these areas in the later period from 1987 to 1989 contained lethal dose levels of HEOD (Newton *et al.* 1991). Therefore, these pesticides may have been at least partly responsible for the widespread decline of barn owls reported during the 1950s and 1960s which was especially pronounced in the eastern counties of England (Prestt 1965). The same may well have been the case in other parts of their range where they were exposed to these chemicals.

Rodents have long been regarded as pests wherever cereal grains or animal feedstuffs are available to them and, despite improved harvesting and storage methods, they still cause problems. Chemical control involves the use of rodenticides, anticoagulants based on the substance cumarin which kills by causing internal haemorrhaging. Warfarin was used widely but rodents developed resistance in many areas prompting the introduction of more toxic 'second generation' rodenticides which include brodifacoum, difenacoum and bromadiolone. Laboratory trials have been conducted in the USA and in Britain in which captive barn owls were fed rodenticide-dosed rodents (Mendenhall and Pank 1980, Newton, Wyllie and Freestone 1990). Brodifacoum was found to be the most toxic: five out of six American, and four out of six British barn owls died after eating the poisoned food. Difenacoum was significantly less toxic and all of the owls survived the trials. However, most suffered short-term haemorrhaging or impaired blood clotting. One of the American birds died after being fed for 10 days with bromadiolone-contaminated rodents.

Difenacoum and brodifacoum came into widespread use during the 1970s and 1980s. An analysis of 145 dead barn owls found in Britain between 1983 and 1989 revealed that 10% of them contained residues of these chemicals (Newton *et al.* 1990). Concentrations of brodifacoum in the liver were generally lower than those in birds that died during the laboratory trials described above. However, the method of collecting these birds was probably highly biased against finding those that had died of rodenticide poisoning. Many of the sample were road accident victims (66%) which obviously had a good chance of being found. Individuals that die of rodenticide poisoning become lethargic and inactive beforehand and it is highly probable that many poisoned owls would have died at their roost sites. These include chimneys, tree holes, hay-barns, roof spaces and other localities where they would most likely not be found. Lethal concentrations of rodenticide might also be lower for wild birds than for captive birds because of their higher activity levels. Sublethal doses were probably also important, causing impaired foraging and death from starvation or increasing bleeding following accidents which could easily happen during moult, when growing feathers are particularly soft and vulnerable. Thus the significance of rodenticide poisoning to wild barn owl populations could have been much greater than indicated by the contamination levels of the birds analysed in this study. There have been numerous reports of individual dead owls being found with clear symptoms of rodenticide poisoning, involving bleeding from the bill and cloaca, but evidence for the wider effects on populations in different areas is inadequate at present and further research is needed.

Barn owls are particularly vulnerable to rodenticide contamination as they often roost and feed around farm buildings, especially in winter, where they can pick up dead or dying rodents. They are likely to be attracted to farms with high rodent populations which are often exactly the places where rodenticides are in use. As few as three poisoned mice would be enough to kill a barn owl (Newton *et al.* 1990) and this number could easily be collected in a single night's hunting.

No evidence of rodenticide poisoning was found in a field trial conducted in New Jersey (Hegdal and Blaskiewicz 1984). None of nine radio-tagged owls living around farms treated with Talon (brodifacoum based) died of poisoning, mainly because they fed away from the farm buildings. However, the trials were conducted in late summer and early autumn when vole populations were high and it would be instructive to repeat these trials during winter, when commensal rodents appear more frequently in the owl diet (Colvin 1984).

A forceful demonstration of the potentially adverse effects of rodenticides was provided by events following the introduction of coumachlor and brodifacoum to control rats in Malaysian oil palm plantations (Duckett 1984). Within 30 months, a population of 40 barn owls was reduced to two, and many were found dead with bleeding from the external openings of the nostrils, a characteristic sign of rodenticide poisoning. Rodents are often serious pests of agricultural crops in the tropics, and the introduction of 'second generation' rodenticides to control them must be a widespread threat to barn owls and other rodent-eating bird and mammal predators in these areas.

16.3 Conservation management

To be successful, a conservation programme for barn owls must ultimately establish a set of environmental conditions that results in viable self-sustaining populations of the owls. Thus, suitable foraging habitat and nest sites must be available at appropriate densities and causes of breeding failure and mortality such as nest disturbance, pesticides or shooting must be brought to tolerable levels. In some cases, solutions may be relatively simple, such as the provision of nest sites, but in most cases, long-term answers will nearly always involve consideration of foraging habitat and other factors.

The need to maintain viable-sized populations rather than just individual pairs of owls cannot be over-emphasised, even though it is difficult on the basis of existing knowledge to specify precisely the characteristics of a viable population. However, a viable population must be one that does not suffer genetic problems, such as those that could arise from progressive inbreeding, and that can maintain itself through the normal variations in environmental factors such as fluctuating prey densities and particularly through periodically catastrophic events such as exceptionally severe winters or droughts. Difficulties will be more severe for populations that are completely isolated than for those that achieve even a small exchange with adjacent populations. The general maxim should probably be 'the larger the better', but in practice it might be realistic to aim for minimum population sizes that fluctuate around 40 to 60 pairs.

The average spacing between pairs within a population must also be considered. In theory, as distances increase, population processes, particularly those involving productivity, could progressively become less efficient, for example, because unmated individuals fail to find each other. At present we know that in some populations breeding birds are highly faithful to their home ranges, even when they lose their mates, so that they may remain unmated for one or more seasons. The dispersal distances of young birds have also been quantified for many populations and in Britain are known to be relatively short (Chapter 13). However, we really have no understanding of the process by which young birds seek out nesting places and mates and there is no scientific basis upon which to judge minimum viable densities or maximum distances between breeding pairs based on the dispersal of the young. The only option, until more information becomes available, is to make a guess that errs towards caution and, from what is known of dispersal distances, an average nest spacing of between 5 and 10 km should avoid any problems, although, for reasons discussed later, higher densities than this will be desirable.

In many parts of continental Europe, young barn owls disperse over much greater distances than those in Britain and it is therefore likely that these populations could survive perfectly well at lower densities.

The management of foraging habitat

The provision of suitable foraging habitat is the most important and yet the most difficult aspect of barn owl conservation because it inevitably involves the encouragement of less intensive agricultural methods or the setting aside of specifically managed areas to increase the food supply. In some regions, existing forms of land use and management continue to provide all that the owls need, and in such areas these needs have to be identified and quantified so that future losses may be avoided. In other areas where the owls are in decline or have disappeared, suitable habitat will have to be created. Individual farmers may be prepared to undertake minor changes to habitat that entail little expense, but there are many areas where substantial, expensive adjustments to land use will be needed if the owl populations are to be retained or reinstated. It is unrealistic to expect farmers to meet these costs or indeed to have the knowledge needed to make the appropriate changes. The process of establishing suitable habitat must therefore have several stages. There must be a detailed understanding of the owls' habitat requirements so that appropriate management prescriptions can be specified. Then, to secure the financial support for farmers to put these prescriptions into operation, a political case has to be made and, lastly, the link must be made to farmers and foresters providing them with appropriate information so that they know what to do. At present, with cutbacks in agricultural production involving the 'set-aside' of farmland throughout all of Europe and North America, there is an opportunity to ensure that at least some of this is done in such a way that it benefits species such as the barn owl with little, if any, extra cost to the taxpayer or additional loss by the farmers.

In this section I will explore only the first of these stages, defining the owls' habitat requirements. At the outset, it must be stressed that there is no single habitat prescription that will fit all barn owl populations in all areas. The owls occur in a great range of environments which differ in many details. In most low altitude farmland in Britain for example, the only suitable habitat for the owls' prey are edge areas such as woodland edges, hedges, fencerows and drainage ditches (Chapter 5). Some areas have few woodlands but extensive networks of drainage ditches, while other areas have many small woods and hedges but hardly any ditches. At higher altitudes, especially in the west, extensive areas of rough grazing and young conifer plantations also provide good habitat. In continental Europe, pastures and meadows that would be of little value in Britain because they are not suitable habitat for field voles support populations of common voles and shrews and can become important foraging areas. In some parts of North America, coastal wetlands offer excellent foraging habitat. Clearly details of habitat are specific to particular areas and many individually designed management plans will be needed. Rather than attempt to investigate the requirements of owls in all possible situations, which would not be feasible in any case as we simply do not have the knowledge, I will aim to identify the main principles involved and illustrate them with a specific example from south Scotland.

The most important point to remember is that habitats must support viable, stable populations of the owls. Thus the quality and quantity of habitat provided must ensure that the population equation resulting from breeding performance and mortality is in balance. Individual pairs within the population can occupy a range of habitat qualities, with some having better areas than others, but over the whole spread of the population the habitat must be good enough to ensure that average production per pair at least offsets the losses from mortality. As both breeding success and mortality vary from year-to-year in relation to food supply and weather, this balance would have to be examined over many years. In practice, it could be extremely difficult to determine the status of a particular population: mortality rates especially are hard to quantify other than by the detailed study of individually ringed birds (Chapter 14). Slow population declines are difficult to detect over short periods of study and seemingly stable populations might only be maintained by the immigration of individuals from nearby better quality areas. Superficial investigations could therefore easily produce completely misleading results, with the risk of eventual failure.

It is relatively easy, by observing the birds, to discover what kind or *quality* of habitat they use for foraging in any particular area, but this on its own is not enough to specify habitat requirements and could, in cases of declining populations, also be highly misleading. The owls might simply be using the best habitat available even though it may be inadequate for their needs. The *quantitative* aspects, that is, how much of particular kinds of habitats are needed, are if anything more important and certainly much more difficult to estimate than simply identifying the nature of the habitats.

The factors responsible for determining habitat *quality* are complex and interdependent, based upon combinations of prey abundance, vegetation structure, prey availability, the owls' foraging behaviour, and various other aspects of the environment. They are also dynamic, varying with season and over longer periods in relation to movements of the prey, cycles of prey abundance and weather.

Some of these relationships have been explored in previous chapters and, in theory, it would be possible to 'construct' an ideal habitat if we understood all of them perfectly. Certainly, on existing knowledge we could go quite a long way towards this, but too many important parts of the puzzle are missing for us to do so with complete confidence or to fullest advantage. For example, we know that field voles prefer areas of mature, damp or moist grassland and we know what kind of food plants they prefer so we could easily describe a type of grassland and a management regimen of grazing or cutting that would produce a good habitat for voles. Briefly, this would involve creating a mosaic of strips or patches of different ages of grassland on, say, a 5- to 8-year rotation so that succession through to scrubland was prevented and mature vegetation offering cover for the voles' runways was adjacent to younger areas containing a higher proportion of the more nutritious grasses and dicotyledenous plants that are preferred as food. In this way, all of the requirements of individual voles could be met in a small area.

However, the factors that control vole densities are poorly understood (we know they are highly complex and dynamic) and we know nothing of the relationship between vole density, habitat structure and the rate at which the owls are able to catch these prey. So, at present, we would be unable to describe the best possible habitat for voles and owls. This might not be important unless we had to achieve the highest densities of owls on limited areas.

The spatial relationships of habitats are always likely to be important. Thus a grassy edge habitat along a ditch might receive an annual influx of wood mice from adjacent cereal or root crops following harvest and temporarily provide a very rich foraging area (Village 1990), whereas a similar edge alongside a closely grazed pasture would not experience this seasonal change. An edge alongside woodland could provide shelter from strong winds, whereas a similar edge alongside a fencerow would offer no protection. Also, the availability of suitable perching posts is likely to be important, especially in winter when the owls may need to conserve energy while hunting (Chapter 6). Thus a whole field of several hectares of mature grassland might be of less value than the same area arranged as wide strips along fencelines bounding a wood (Fig. 16.4).

Quantitative aspects of habitat, that is the density of suitable foraging area per hectare, is particularly important in summer when the owls are feeding their young and making repeated foraging trips back and forth to the nest. The greater the density of the food supply around them, the easier it should be for them to capture prey rapidly and minimise the energy and time costs of transporting it to the nest. This has been shown to have profound effects on the production of offspring (Fig. 12.12, p. 182). Also, the higher the prey density, the lower should be the rate at which prey are depleted by the owls. This has never been calculated for barn owls but other vole predators have been shown to consume significant proportions of the available prey (Fitzgerald 1977, Goszcynski 1977, Southern and Lowe 1982, Erlinge *et al.* 1983). Young voles do not mature and reach a size that is taken by the owls until mid to late summer, so that for some of the chick-rearing period the owls depend upon a diminishing stock of adult voles (Chapter 5).

The continuity of habitat at suitable densities over large areas is also important. An isolated patch, large enough to support a pair in summer, may not fulfil their requirements in winter when they forage over much larger areas (Chapter 7). Also, isolated patches, say 20 or 30 km apart, may simply not be discovered rapidly enough by potential settlers to be occupied consistently.

Barn owls will always have to fit into whatever happens to be the prevailing agricultural and forestry landscape of any particular region, with perhaps the possibility of some relatively minor modifications of

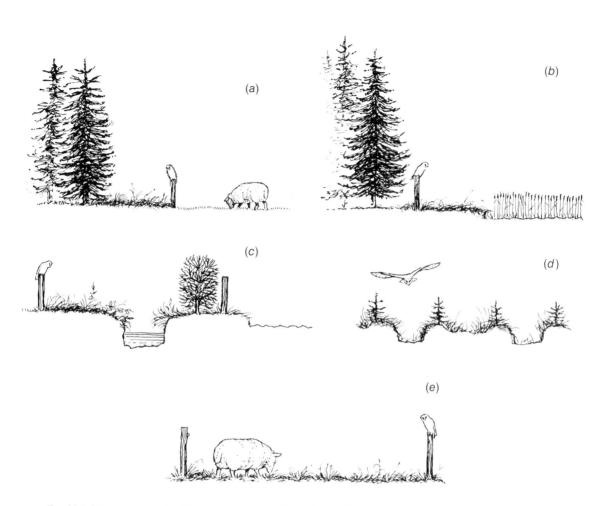

Fig. 16.4. Diagrammatic representation of suitable small mammal habitats and barn owl foraging areas in farmland. (a) Grass strips, 3–5 m wide, along woodland edges, with fenceline excluding livestock or pastoral areas. (b) In arable areas, the grass strip can be established on the field side of the fence, allowing easier management. (c) Grass strips along streams, ditches and hedgerows, with fencelines excluding stock and providing perches. (d) Strips and small patches of young woodland. (e) Areas of damp, lightly grazed rough grassland. (Drawings: A. Baxter Cooper.)

habitat to accommodate them. Detailed prescriptions of habitat management for them based upon the above principles will therefore have to be reached independently for different areas. However, it is instructive to demonstrate how this might be done by working through the specific example of my study area in south Scotland. The low altitude enclosed farmland section of this area is quite similar in land use, habitat structure and climate to much of western Britain. Devoted primarily to dairy and livestock production, it consists mostly of short-rotation pastures, silage and hay, interspersed with small woods, hedges, fencerows and ditches.

The objective was to specify the type and quantity of habitat that could be maintained in this environment to sustain a viable population of the owls. The various components of the process leading to this have been detailed in several of the preceding chapters so an abbreviated form bringing them all together is given here. The first step was to discover what prey the owls consumed and where they caught it. By the analysis of pellets over 13 years and four complete vole cycles, it was established that field voles were the dominant prey but that during the decline phase of the vole cycle the owls switched to common shrews and wood mice (Chapter 3). These alternative prey were therefore also important, as were the habitats that supported them. Then, by using radio-telemetry it was established that the owls caught most of their prey from grassy strips along woodland edges. These strips varied in width but averaged about 3 to 5 m and were always separated from adjacent fields by fencerows. The owls hunted both in flight and from perched positions on the fence posts. The availability of these perching places was probably an essential component of the system, enabling the birds to use the perched hunting technique in winter when heat and energy conservation was important (Chapter 6). Perching may also have been important during strong winds and rain but this level of detail is difficult to obtain from radio-telemetry on dark winter nights.

The grassy woodland edges supported dense populations of field voles but the trees must also have produced sheltered foraging conditions. The owls may therefore have been able to continue foraging in this habitat when conditions in more open landscapes would have been difficult. The woodland may also have provided a 'reservoir' of prey, maintaining densities at high levels in the grass strips by replacing individuals removed by the owls (Chapter 7). The grass strips did not need to be managed, at least in the short term, possibly because intense grazing and browsing by roe deer prevented the regeneration of scrub.

From the radio-telemetry work, it was established that the average pair of owls had a home range of 3.2 km² during the breeding season, equivalent to a radius of 1 km around the nest. During the winter, they ranged more widely, up to 5 km from the nest (Chapter 7).

From 1979 to 1991, a detailed study was made of the owl population dynamics in this habitat; productivity, mortality and the rate of recruitment to the breeding population were quantified. A simple population model was constructed and from this it was shown that a long-term average productivity of about 3.2 young per pair would be needed to maintain the population (Taylor and Massheder 1992).

The characteristics of the habitat within the study area and around the owl nests were quantified from ground survey. A strong positive relationship was found between the birds' breeding success and the length of woodland edge within the 1 km radius of their nest. To achieve the required productivity of 3.2 young per pair, about 8 to 10 km of woodland edge within the 1 km radius was needed (Chapter 12).

This then suggests that a stable population could be maintained by providing numerous small farm woodlands, with a 3 to 5 m wide grass strip edge, bounded by fencerows from adjacent crops such that, on average, the owls had 8 to 10 km of this edge within a 1 km radius around suitable nest sites. To this formula has to be added a slight modification to cater for the birds' behaviour in winter when they utilised larger areas. Isolated units of 1 km radius probably would be inadequate. Larger units of about 5 to 10 km radius having habitat of this density, should be aimed for. This could accommodate several pairs of breeding owls and also fulfil their winter requirements.

The existing habitat within the study area could be improved greatly as there are many woodlands that either lack grass edges or have very narrow ones. Through time, as the woods are managed, the landowners could be encouraged to reconstruct them, leaving edges of appropriate width. Also there were many patches of habitat that have a high density of suitable woodland edge but no nest sites, so that, even with no change in habitat, considerable improvement could be achieved by the judicious placement of nest boxes in these areas. Equally there are good traditional nest sites in areas that have become poor habitat that consistently attract birds to breed, but, when they do so, their production is far short of that required to replace losses. The potential of these birds is in effect wasted and it would really be better to prevent access to these sites so that the birds were encouraged to breed more successfully in the higher quality habitat where new nest sites could be provided. The old nest sites could be opened up again if and when the habitat around them is improved. Such an approach could only be taken after very serious consideration based on a great deal of knowledge and appropriate permission.

There is no reason why the above habitat prescription, based on numerous small farm woodlands, should not work in many parts of the barn owls' range in Britain. Indeed, in these times when farmers are being forced to diversify, it may be entirely compatible with future changes in land use, brought about not for conservation reasons but simply to enable farming and rural communities to survive.

The process of quantifying the owls' requirements described above may seem complex and many may believe that 13 years is a long time to wait for an answer. Certainly it would be inappropriate to do nothing at all until a definitive answer is produced but equally we must avoid the situation whereby government (tax payers) and farmers are persuaded to invest a great deal of money, land and effort into providing habitat only to find that it is inadequate. Long-term research should go hand in hand with immediate conservation efforts but in such a way that the argument is left open and flexible until reliable final recommendations are available.

Management of nest sites

It seems likely that most barn owl conservation programmes in Europe and North America will eventually need the long-term management of nest sites, involving the maintenance of traditional sites and the provision of new ones. There is evidence from many areas that population declines can be prevented and even reversed by nest management (Braaksma and de Briujn 1976, Juillard and Beuret 1983, Colvin 1984, Regisser 1991). However, there are also regions where agricultural intensification has progressed so far that there is now almost no suitable foraging habitat and the provision of nesting sites is pointless without first reinstating the habitat.

Nest site maintenance is usually expensive and labour intensive. Many can be lost each year through natural events and unwittingly through human activity and so, to prevent a gradual loss, sites should be checked each year in late winter before the birds start settling in. Nest boxes should be emptied of pellet debris, again preferably every year and, as no nest box lasts forever, a regular cycle of replacement is needed. It is a commitment that probably will have to be continued for as long as we want to have barn owls around us.

Nest boxes can be used in more subtle ways than simply for the replacement of lost nest sites. In most areas there are patches of good habitat that remain unused by the owls because they lack nest sites. By identifying these patches and ensuring that they all have a high density of nest boxes in them which are used by the owls, the productivity of populations can be increased. Making the best use of existing habitat in this way may be all that is needed to maintain some populations and for others it can be a valuable initial step to slow down the rate of decline while attempts are made to improve the habitat more widely.

The provision of nest boxes can produce spectacular increases in barn owl numbers in areas where there is exceptionally good foraging habitat but few nest sites. In south Scotland, a small group of researchers (Andy Dowell and Geoff Shaw of the Forestry Commission and myself) set up a nest box trial in an extensive area of established conifer plantations. The habitat was varied, with newly planted and clearfelled compartments in a mosaic of closed canopy forest and grassy rides. At the start there were only about six or seven nest sites and four to six pairs of barn owls, although the surrounding farmland had long supported a dense population. Nest boxes in the form of 20 gallon plastic drums with entrance holes cut in the side were strapped to spruce trees at heights of about 4 to 5 m, at 57 sites. Two drums were placed at each, so that if tawny owls which were already common in the forest occupied one, the other was still available for the barn owls. Within 4 years, the

Fig. 16.5. Nest boxes installed in oil palm plantations (right) and rice growing areas (left) in Malaysia have resulted in dramatic increases in the number of breeding pairs. (Photographs: Iain R. Taylor.)

barn owl population rose from 6 to 29 pairs and then fluctuated around 20 to 30 pairs in response to the field vole cycle. During this build up there was no decline in the numbers nesting in the surrounding farmland and the initial influx was mainly of young birds reared on the farmland. It seems likely that before placement of the boxes the plantations may have supported a substantial non-breeding population of the owls (Taylor, Dowell and Shaw 1992).

Another much more striking example is provided by nest box programmes in the Malaysian oil palm plantations. This crop has expanded from 54 638 ha in 1960 to 1 474 135 ha in 1987, unfortunately through the clearfelling of rainforest. Barn owls were first recorded nesting in Malaysia in 1969, in the attic of John Duckett's house set in an oil palm estate (Wells 1972). Further investigation showed that others were nesting in the hollow stumps of former rainforest trees but Duckett (1976) noted that there were many non-breeding adults in the plantations. Graham Lenton (1980) demonstrated that these birds would breed when nest boxes were provided (Fig. 16.5) and this raised the

Fig. 16.6. Disused silos such as this one in New Jersey are often used by roosting barn owls and are good sites for the installation of nest boxes. (Photograph: Iain R. Taylor.)

possibility of using barn owls as biological control agents to reduce crop losses caused by rats. In an extensive trial, 200 boxes were erected in an area of 1000 ha of mature palm plantations in the Kok Foh Estate, south east of Kuala Lumpur at the start of 1988. These were taken up quickly and by December 1989, only 20 months later, 95% were occupied by breeding barn owls (Duckett 1991). At about one pair per 5 ha this must surely be the densest barn owl population anywhere and the total plantation area of Malaysia presumably has the potential to support almost 300 000 pairs.

The coastal marshlands of the eastern United States, around Delaware and Chesapeake Bays probably offer the best and most extensive areas of barn owl foraging habitat in North America but nest sites within these marshes are few. Many owls resort to nesting on duck shooting blinds anchored offshore, often sharing them with ospreys (Reese 1972). Barn owls in this area will commonly forage up to 3 km from their nests and sometimes up to 5 km (Hegdal and Blaskiewicz 1984). Thus they are able to commute out into the marshes from farms along the edges and nest boxes sited in the barns of these farms along one part of the New Jersey coast in the 1980s have resulted in the establishment of many breeding pairs (Bruce Colvin, personal communication). There must be the potential to increase the population of owls on this whole coastline by many hundreds simply by providing nest boxes and, as most of the marshland habitat is now protected, the long-term future of the owls could easily be secured (Fig. 16.6).

The above are somewhat unusual examples but nevertheless demonstrate the potential of nest site management. More typically there are many areas, especially in continental Europe, where nest site loss was suspected to be the main cause of population declines and where a combination of restoring old sites and providing new ones has resulted in substantial population increases. One of the earliest of these projects was in the province of Gelderland, Holland, where, between 1968 and 1975, 29 nest boxes were erected of which 15 were used for breeding (Braaksma and de Bruijn 1976). Indeed, Dutch farmers have a long history of encouraging barn owls through the provision of specially ornate entrances (Fig. 16.7) allowing the birds access to breeding areas inside barns. However the availability of suitable nesting places throughout Holland has declined progressively through the modernisation of buildings and the wiring up of dovecots and church towers to exclude pigeons. Starting in 1976, a conservation programme was established in Friesland in which, among other measures, nest boxes were provided and access re-established to blocked off sites. From 18 pairs in 1976 the population increased to almost 200 pairs by 1990 (de Jong 1991).

Another large-scale nest management scheme has been undertaken in the Haute-Rhin region of France where between 1978 and 1990, 215 nest boxes were installed mostly in church towers and barns

Fig. 16.7. Farmers in Friesland, Holland have traditionally installed a specially ornate board ('uilebord', owl-board) with a hole ('uilegat', owl-hole) at the gable end of their barns, allowing barn owls to enter. (Photograph: Johan de Jong.)

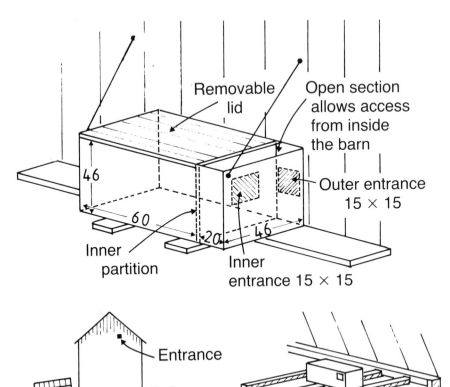

Fig. 16.8. Some barn owl nest box designs.

(a) A rectangular box made of good quality 16 mm plywood or softwood, suitably treated with non-toxic wood preservative. The box may be attached, inside barns, to the gable-end wall with a hole cut to allow access from the outside. An open section of the lid allows access to the owls from inside the barn and enables the young to climb onto the lid to exercise, reducing the chance that they will leave the barn prematurely. Where mammalian predators are a problem, or where the owls cannot be allowed access within the barn for hygiene reasons, this opening can be dispensed with so that entry to the box is possible only through the barn wall. The box can be also be attached to rafters or roof-ties with access only from within the barn. Measurements are in centimetres and the drawings are not to scale.

(b) A simple rectangular box, with removable lid, simulating natural vertical tree cavities. Can be attached by nylon rope to tree trunks.

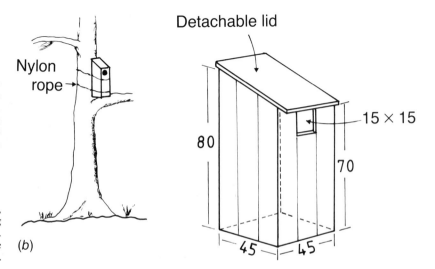

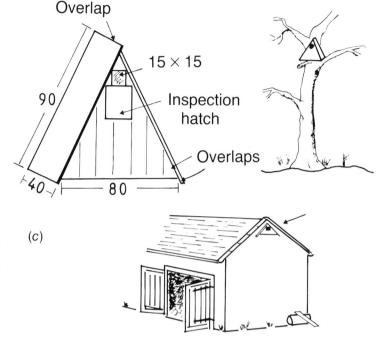

(c) A more elaborate A-shaped design made of exterior quality plywood or treated soft woods, simulating traditional dovecotes, which gives good waterproofing and can be attached to trees or to the outside gable ends of farm buildings. Measurements are in centimetres.

(Regisser 1991). The population of owls increased from 15 known breeding pairs in 1976 through a series of fluctuations to a total of 90 pairs in 1990.

The most appropriate dimensions, designs and positioning of nest boxes will depend mainly upon local circumstances and the subspecies of the owl involved (Fig. 16.8). Boxes can be attached to trees but these can sometimes attract potential competitors and predators. For example, in southern Scotland I attached 10 wooden nest boxes to farmland trees in 1981 and, although initially barn owls bred in four, eventually all were occupied by jackdaws. The owls often evicted jackdaws from chimney nest sites and presumably could have done so from these boxes. The main problem was that the jackdaws being much more numerous than the owls, usually were the first to use the boxes and when they did they filled all but a small section with sticks and sheeps' wool such that the remaining cavity was too small for the owls. Problems such as this can be reduced by visiting the boxes immediately after the breeding season and emptying them of any jackdaw nests so that they are available overwinter to any prospecting owls. As the owls occupy the sites slightly earlier in the year than jackdaws and are able to fend them off, this should solve the problem but, of course, takes time. Alternatively, in particularly vulnerable areas, the boxes can be placed inside buildings and there is the additional advantage that by doing this cheaper, non-waterproof materials and designs can be used.

In most circumstances a simple rectangular-shaped wooden box is all that is needed. Size is a compromise between the minimum accepted by the owls and the costs involved. For both European and North American birds a floor space of approximately $0.4 \, m^2$ and a height of 40 cm is about ideal, although boxes considerably smaller than this will sometimes be accepted. The entrance hole should be on one of the sides, towards the top; this helps prevent young birds leaving prematurely and also means that pellet debris can accumulate without blocking the hole. A removable lid is useful for cleaning and inspecting the nest contents. The most suitable shape depends upon where the box is to be installed. A rectangular design (about 130 cm $\times$ 40 cm) is most stable when the box is attached to the side of a barn wall (Fig. 16.8), otherwise shape seems unimportant. The box floor should be lined to about 5 to 10 cm deep with fine wood chips or shavings. Horticultural peat should not be used.

Installation sites should offer security from predators and human vandalism, easy access for the owls at all times uninhibited by human activity, and freedom from frequent and prolonged disturbance especially in the early spring when the birds are in the initial stages of courtship and nest site selection. Particular birds may become habituated to human activity and noise with time but new birds tend to be wary. Barns, haysheds and lofts that are not in constant use and preferably away from centres of activity on the farm are the most appropriate sites. Boxes are best placed as high as possible on rafters with access through open sides or doors, ventilation slits, or specially cut entrances. Heights of at least 6 m are preferred by North American birds (Colvin 1984) whereas in Scotland, where there are no nest predators, boxes installed at only 2 m have been used but only in buildings that were hardly every visited by humans. The birds can be allowed free access within the barn or the box can be attached directly to the inside of a gable wall with access only to the outside (Figs. 16.9–16.11) which further reduces the possibility of predation or human disturbance but may in some areas increase the risk of other species using the boxes or of the young leaving prematurely. This arrangement is most suitable if a farmer objects to pellets or droppings within the barn.

A variation of this design has been used successfully in church towers that have been wired up to keep pigeons out. In this situation the box can be placed inside, close to the grill with a tunnel linking the box to a small opening cut in the wire mesh. Thus the owls can use this

Fig. 16.9. Entrance hole to a nest box installed to the gable end of a barn by Bruce Colvin, New Jersey. (Photograph: Iain R. Taylor.)

Fig. 16.10. A nest box attached to the gable end of a hayshed, seen from inside, south Scotland. (Photograph: Iain R. Taylor.)

traditional and probably preferred type of site whilst pigeons are still excluded from the tower (de Jong 1983).

Where there is no opportunity of placing boxes inside buildings, it is still possible to attach waterproof boxes to the outside, an idea that has

Fig. 16.11. A specially designed nest box used for the capture of the adult birds, south Scotland. The box has a long entrance tunnel and when the chicks are about 4–6 weeks old, a box trap can be slipped within the tunnel which closes when the adults enter with food. This has been found not to damage the birds in any way nor to cause desertions, but it must be remembered that appropriate government or state permits are needed to visit nests and to capture the birds.

long been used for kestrels. In Scotland, I have used an A-shaped box for this purpose, modelled after some of the simpler traditional dovecots (Fig. 16.8). The front and back are made of exterior quality plywood, the floor of 16 mm thick softwood and the roof of exterior plywood or overlapping slats of European larch (*Larix decidua*) which provides cheap and durable waterproofing.

Boxes used in Malaysia have been free-standing, supported on 4.5 m high wooden posts sited under the shade of palm fronds (Fig. 16.5, Duckett 1991). Similar designs have been used for some time in Europe for kestrels, especially in reclaimed polder areas of Holland (Cavé 1968). The usefulness of such boxes for barn owls in Europe is probably limited by costs, as a slightly larger structure is needed than for kestrels, and by their probable attractiveness to other species, such as jackdaws.

Winter feeding

Most barn owl mortality, especially of breeding-age birds, occurs during winter and the main cause seems to be starvation (Chapter 14). In the Scottish area, mortality rates were highest during winters of low vole abundance and exceptionally severe weather and overwinter mortality was the main determinant of yearly changes in breeding numbers (Chapter 15). This suggests that the provision of food during winter should enhance survival and reduce population declines. Barn owls will certainly identify and consume dead prey left at roost sites (Chapter 7). This is simply an extension of their normal behaviour in which food items are cached around the nest in summer and sometimes at roost sites in winter, for consumption at a later date.

Widespread continuous food provisioning is probably undesirable as it might result in dependency but, as part of a comprehensive conservation strategy for populations that have become reduced to very low levels, it may have a useful role. It is unrealistic to expect that food could be provided throughout the whole winter and the obvious approach would be to provide it at times when weather conditions are likely to reduce the prey capture rate: during periods of continuous strong winds, sleet and heavy rain and when there is prolonged deep snow cover with extremely low temperatures. In cold conditions, food items quickly freeze up and therefore have to be provided daily, but it is important not to scare birds out of their roosts when doing so as this may substantially increase their energy expenditure. Items can be left on window ledges or other exits around the nest building or roost in the evening or at feeding stations which can be set up at other buildings used by the birds. There is, however, a risk that by feeding owls in this way they might become trained to search for dead items in buildings more generally which could bring them into contact with rodenticide-poisoned rodents. In cereal growing areas where this is most likely to happen, it is important to make farmers aware of the problems so that rodenticides are used with care and preferably not during severe winter weather.

Winter feeding has been part of the package of conservation measures associated with the successful recovery of the barn owl population of Friesland, Holland (de Jong 1991). Food was provided directly in barns, but also indirectly by encouraging wild rodent populations to concentrate around grain laid out in barns used by the owls.

Large populations of owls in areas of good habitat should have the resilience to increase rapidly even after severe reductions (Chapter 15) so winter feeding is unnecessary.

Control of rodenticide use

In Europe and North America, commensal rodents will continue to be regarded as pests and their control will be deemed necessary; it is unrealistic to expect that farmers will completely forego the use of rodenticides. There is a good case for further research into the effects of these poisons and the ways in which they can be used without harming barn owls. Increased hygiene on farms might help in the long term, by minimising the amount of food available to rodent pests and hence reducing their populations without the use of chemicals. Also, the improvement of grassland habitats might increase the natural prey for barn owls away from buildings and decrease the chances that they will take the poisoned rodents. However, neither of these measures could be guaranteed to work in all places at all times and additional measures are desirable.

Rodenticides such as difenacoum which are less toxic to the owls should be used in preference to more harmful substances. Rodent control would be more effective and probably less harmful to the owls if done mainly in autumn and early winter when most rodents move from the fields into farmsteads (Huson and Rennison 1981) and before the onset of severe weather that might encourage more owls to forage around buildings. Also, if periods of about 10 to 14 days of rodenticide application using difenacoum were alternated with similar periods of no poisoning, this would reduce the chances of individual owls accumulating lethal doses and allow them to rid themselves of any poison they had picked up. The owls should, if possible, be excluded from any buildings in which poison is being administered and the exit of poisoned rodents should also be minimised.

It would be preferable for poisoning to be abandoned during periods of very severe weather when the owls would be most vulnerable but this might not always be possible. In some circumstances, with some forethought and a little work, uncontaminated food could be made available to the owls in conspicuous places used by them if rodenticides have to be used at such times.

Measures such as these would have to be undertaken by all farmers over large areas, as individual owls will nearly always forage around several farmsteads during winter and the objective should also be to protect viable populations of the owls rather than just individuals.

16.4 Monitoring barn owl populations

The purpose of monitoring is to detect change in the status of species, usually in their breeding numbers but sometimes also in their distribution, so that the effects, if any, of human activity may be determined. With so many barn owl populations in Europe and North America being in a vulnerable condition, some kind of monitoring of the situation is highly desirable. However, assessing the abundance of barn owls is an immensely difficult task for many reasons not the least of which is their unobtrusive, usually nocturnal behaviour which makes them hard to locate. Ambitious monitoring plans may prove difficult in practice and it is therefore important at the outset to appreciate all the potential problems. There is also no point undertaking a costly and time consuming monitoring programme without obtaining some measure of the reliability of the figures obtained. Errors, or sources of unreliability, can enter at many points of the exercise and it is important to identify them so that they can be minimised and, at the end, quantified. Failure to do this may give rise to misleading results which not only waste valuable resources but can also be severely damaging to the original conservation objective. At its worst a bad monitoring programme could suggest an improvement or no change in a species' status when in reality a decline is taking place.

Monitoring barn owl populations could be done at local, national or international levels but the larger the area the greater the problems. Rarely is it possible to count all of the individuals or pairs directly, other than at a local level, so for larger areas a sampling programme must be devised and this is where many of the difficulties arise. Monitoring by sampling only ever gives an *estimate* of the species' mean density or abundance and the reliability of that estimate is given by the standard error of the mean value obtained. Ecologists usually express this in terms of confidence limits, that is, an upper and lower limit around the estimated mean within which the real or true mean can be said to lie with a specified degree of certainty. The practical significance of this is that we can only be sure that an apparent change in the abundance of the species is real, and not simply a result of sampling, if the differences in estimated mean densities exceed the magnitude of their confidence intervals.

For species living at reasonably high and regular densities in a uniform environment, this poses few problems but barn owls usually occur at overall low density and are very unevenly distributed, presumably in relation to the spacing of suitable nests and habitat. The implications of this are best illustrated by a practical example. In south Scotland, a group of workers all with long experience of barn owls attempted to estimate the total number of breeding pairs in an area of 6300 km^2 (Taylor *et al.* 1988). Sample units measuring 10×10 km were used and these were searched exhaustively for barn owl nests. This revealed densities ranging from 0 to 13 pairs per square. However, even with a sample of 22 squares representing 36.5% of the total area, the 95% confidence intervals were still 27% of the estimated mean density. This indicates that we would only be able to detect a real change in the owl population density with reasonable statistical certainty if that change was an increase or decrease greater than 27% of the estimated original density. Stratified sampling, a method commonly used to improve the reliability of density estimates could not be employed without a massive ground survey because variations in the owl population density were piecemeal and not easily arranged into broad, readily identifiable zones.

Detecting change in barn owl populations by comparing estimates made at different times is clearly very difficult, so, rather than relying upon individual estimates of density spaced perhaps 5 or 10 years apart, it would be better to sample more frequently and analyse longer runs of data for trends. Here, the barn owl throws us another problem. Some populations, such as the one studied in Utah by Carl Marti (1988), vary little from year to year but most and perhaps all European populations vary greatly in response to vole cycles and exceptionally severe winter weather. Any man-induced change in status would be superimposed upon these natural fluctuations. With a 3- to 4-year periodicity and variations around this for the vole cycle and irregular, unpredictable weather effects, there is no option other than to sample continuously, every year to have even a modest chance of detecting a long-term increase or decrease. Sampling at lower frequency, say 2- or 3-year sample periods every 10 years is likely to produce results that cannot be interpreted and could be highly misleading.

Continuous monitoring over many years has been carried out for several populations in Europe and North America. However, in nearly all cases this has been accompanied by some conservation work, especially the protection of nest sites, so any trends shown in these populations may not be representative of wider areas. To be representative, a sampling programme should be based upon randomly placed sample units. This could be achieved by using map co-ordinates generated from random number tables to define a series of 10 km sample squares. Ideally, a new series should be selected each year but in practice, because of the time needed initially to locate potential nest sites, the temptation would be to re-use the same squares. This would be equivalent to fixed position quadrats, frequently used in ecology, but it is essential that the observer's presence in such quadrats does not alter the events within them, so that they remain representative. With barn owls this important requirement is unlikely to be met. Few observers would be prepared to stand back and watch their barn owls disappear through the loss of nest sites, for example, and sample quadrats could very quickly become unrepresentative. The opposite could also occur if observers did nothing in the sample squares yet all around the population was increasing as a result of conservation measures. This source of error alone could completely invalidate any monitoring programme. Meaningful ecological units for the barn owls would have to be identified and such units treated separately in the sampling procedure.

Attempts to monitor on a large scale, say the size of a whole country such as Britain or England, not only presents these problems of sampling but additionally may lead to difficulties in interpretation of results. Such large areas usually have several climatic zones and regions of very different land use and the owl populations in them could behave in different ways. An increase in one area could be obscured by a decrease in others if the whole is treated as a single unit.

All of this suggests that reliable monitoring may only be possible at a more local level, under tightly controlled conditions to answer well-defined questions. However, even here there are pitfalls for the unwary. For example, an area might be set up to test the effects of particular conservation measures such as changes in land use or the provision of nest sites, but the results of this will be uninterpretable in many cases unless a control area is set up as well. The control area is one that is matched in every respect to the trial or experimental area except that no conservation measures are undertaken; the need for it arises from the fact that many environmental variables apart from the conservation measures could affect barn owl numbers. An increase in numbers in the trial area could, for example, be the result of the recovery from exceptionally severe winters or a run of unusually high vole densities rather than the conservation measures undertaken. Control areas help to ensure that the correct interpretation is made, but, of course, double the effort involved.

The whole question of monitoring needs very careful consideration. A general formulae is inappropriate because the problems will vary in their nature and extent from one population to the next and according to the questions being asked. It is highly desirable that a professional ecologist who understands sampling procedures and statistics be involved at least in the planning of any monitoring programme.

16.5 Summary

Barn owl populations have undergone widespread and serious declines over much of Europe and North America since about the 1940s, related primarily to the intensification of agriculture. This has reduced the amount and quality of habitats containing populations of small mammal prey. Pesticides, especially Aldrin-Dieldrin and rodenticides, have almost certainly also contributed to declines in many areas, as has the loss of nest sites in traditional farm buildings. Conservation programmes for barn owls must concentrate on providing appropriate habitats at adequate densities and on the maintenance of nest sites. Strict controls on the use of rodenticides are also needed.

17 Conclusions

I t seems fairly certain that barn owls originated in warm, tropical or sub-tropical savannahs, open woodlands and semi-arid habitats. The relatively poor insulating qualities of their plumage, giving them a thermal neutral zone of 25 to 33 °C is certainly not characteristic of a species adapted to cold climates. Traditionally, barn owls have been typified as long-winged, long-legged birds of prey that hunt small mammals over open habitats and, while this general description is probably fair for most of the many subspecies, it is also clear that it is not entirely accurate for all subspecies. Through time, the birds have spread widely into most continents and onto many distant islands, and, by isolation and changes in selection pressures, there has been considerable divergence: subspecies differ in coloration, in body size and in the relative proportions of wing, tail and leg lengths in ways that seem to be adapted to the different habitats in which they live.

I have tried to present as broad a picture as possible of barn owl behaviour and ecology including the variations among subspecies. However, it must be remembered that most knowledge comes from studies of the subspecies in Europe and North America; we have some information for some others in Africa, Malaysia, Australia and the Galapagos Islands but for most we know nothing. We have been able to elucidate the main themes and many details of the owls' ecology, but many exciting variations undoubtedly remain to be discovered, especially for tropical subspecies living in more forested habitats.

Predation, food supply and the evolution of life history traits

In different chapters of the book I have described various aspects of the barn owls' ecology: when and where they breed, how many eggs they produce, the size of their eggs, the growth patterns of the young, moult, dispersal patterns and how mortality varies with age and other factors. All of these could be described as 'life history traits' and, at each stage, I have attempted to explore how they are linked to the barn owl's particular mode of predation and to the variations in food supply available to the birds.

In nature these traits are, of course, not independent of each other; they interact and the outcome of these interactions is what determines the fitness of individual animals, in an evolutionary sense. Interactions are usually in the form of 'trade-offs'; that is, potential changes in one trait may be constrained by the effects on others. For example, a bird may rear more young in one brood by laying more eggs, but in doing so it might lower its body condition and thereby reduce its chances of surviving to rear a second brood. The result would be that fewer young are produced in total than if it had laid fewer eggs to begin with in the first clutch. The consideration of all traits together, and in relation to the birds' mode of predation and food supply, is thus a highly constructive way of arriving at a better understanding of the significance of individual traits and of how they have evolved. In this discussion I will try to bring together the relevant threads that have been developed somewhat independently in the preceding chapters in an attempt to reach a better understanding of the barn owl's life history.

Over most of their range barn owls are specialist predators of small mammals and in most areas they depend mainly upon rodents, particularly voles, rats and mice (Fig. 17.1). Often only a single species predominates in their diet, such as field voles in Britain, meadow voles in the eastern USA, multimammate rats in Africa and house mice on many islands. There is considerable evidence that such species are captured

Fig. 17.1. Barn owls are specialist predators and, throughout most of Europe, depend very heavily on voles. This female returns to the nest with a large field vole.

preferentially and that for reasons of body mass, behaviour and habitat they are the most profitable prey species for barn owls in their particular areas. All of these prey undergo marked fluctuations in abundance, with some notable exceptions, such that periodically they reach very high densities and at such times barn owls may specialise almost entirely on these individual species.

The owls widen their diet to take in other prey mostly when the abundance of the preferred, most profitable species declines. Their behaviour thus follows the general predictions of the optimal foraging theory (Charnov 1976). Switching to other prey and broadening the diet can take place seasonally, cyclically, or more irregularly according to prey abundance, and more generalist diets also occur in less productive environments such as semi-arid areas. There seems to be considerable variation among subspecies in the range and variety of alternative prey but in most situations this is restricted to a limited range of other small mammals. In much of Europe, for example, shrews are the main alternatives. Only in very special situations are other taxonomic groups included: birds become important only where they occur at high densities in communal roosts, feeding flocks or nesting colonies within the hunting ranges of individual owls. Insects, mostly large grasshoppers, crickets and beetles, are significant only in Mediterranean climates, especially on islands, and earthworms seem not to be taken at all.

The barn owl is thus a much more specialist and less flexible predator than most others that depend mainly on small mammals. Most forest owls such as the tawny owl can switch more to song birds, insects and earthworms (Southern 1954) when small mammals decline, as can some diurnal raptors such as the European kestrel (Village 1990). Even the short-eared owl, normally thought of as an extreme vole specialist can switch much more easily to birds in winter, when voles are less available, than can barn owls (Fairley 1966, Schmidt and Szlivka 1968, Glue 1977), although this may arise at least partly through seasonal changes of distribution and habitat. Barn owls may not extend their diet as widely as other species either because they have no need to or because in most habitats they cannot. The observation that they tend to do so when the opportunity arises suggests that the latter is generally the more likely explanation. Why they should avoid earthworms is unclear, especially in parts of Europe where kestrels living in exactly the same habitats are able to add them to their diet when voles are scarce. Barn owls hunting from perches along the edges of pastures must frequently encounter earthworms and there seems no reason why their morphological adaptations should prevent the capture of these prey. Presumably, they are not recognised as prey or are avoided for other reasons.

By switching to other prey, the owls do not compensate for the loss of their preferred prey species. This is implicit in the argument that the preferred species are also the most profitable. In the Scottish area, for example, pairs nesting in conifer plantations preyed more on common shrews when field voles declined but in doing so their average prey weight was reduced by 38% comparing the years of highest and lowest vole abundances. Also, as the shrews declined at the same time as the voles, perhaps as a consequence of shifts in predation pressure, the overall prey capture rate of the owls must have declined, although this was not measured. Simultaneous declines in various small mammal species have been found in quite a number of northern areas but how widespread the phenomenon might be in other parts of the world is unknown.

Generally, birds that depend upon markedly unpredictable food supplies respond either by remaining where they are and, by diversifying their diet, attempting to survive through periods of shortage although they may be unable to breed at such times, or by dispersing widely in an effort to find a new area of high food abundance. Once an alternative area has been discovered, the improved food supply there allows successful breeding, better survival and rapid population growth. Such species have been termed 'r-selected', implying that they have evolved under density-independent conditions, that is, under conditions where their breeding and survival are not significantly influenced by the density of conspecifics (MacArthur and Wilson 1967). However, the r-selection concept has recently attracted serious criticism (e.g. Stearns 1992) for a variety of reasons. Whatever, some of the barn owl's characteristics are typical of what have been called r-selected species and some are not.

In the Scottish population, the adult owls were almost totally faithful to their nest sites even when vole populations crashed to very low levels and most young birds also settled close to their natal areas regardless of vole abundance. They did not disperse away from the area when food became scarce. As a result of this and their specialised food habits, many died of starvation. Mortality rates varied greatly from year-to-year in both age groups, reaching levels as high as 55% among the breeding adults in the years of lowest vole abundance. Published information on movements for other barn owl populations with fluctuating food supplies is patchy and not entirely consistent. The analysis of national ringing recoveries suggests that barn owls throughout Britain behave in much the same way as those in south Scotland (Percival 1990), and breeding adults in Burgundy seem to be just as faithful to their nest sites as the Scottish birds (Baudvin 1986). However, in central Europe generally, juveniles disperse more widely when voles are scarce than when they are abundant, thus differing from British birds. Birds ringed as adults in Germany also moved more when voles declined but, as no information was given on the status of such individuals at ringing, it is not known whether they were established breeders or adult non-breeders (Bairlien 1985). There are many anecdotal accounts of extensive movements, accompanied by mortality, of Australian barn owls following the collapse of house mouse populations but whether the dispersal involves both adults and juveniles is unclear (Fleay 1968, Schodde and Mason 1980, Hollands 1991). More information of a more precise nature is needed for populations outside Europe before any firm general conclusions can be reached on the nature of barn owl movements in response to fluctuating food supplies.

The absence of movement in breeding birds, with a high degree of nest site fidelity even when prey abundance declines, might be expected if suitable nest sites in good quality habitat are scarce. In this case the gains from remaining, despite the increased chances of mortality, would outweigh those from moving with the risks of failing to find either a suitable nest site or areas of high prey abundance. Male Tengmalm's owls tend to remain and defend their nest sites when voles decline, as alternative nest sites are in short supply (Kopimäki 1987), and the same argument might hold for European barn owls.

Lower nest site fidelity and an increased tendency to move among breeding adults might be selected for in environments where the mortality risks from remaining are exceptionally high because food supplies fall to much lower levels, or where alternative nest sites in other areas are abundant such that the risks of failing to find an alternative nest cavity are not so high. This might indeed be the case for barn owls in some of the woodland or semi-arid habitats of Australia and other areas. It is thus possible that different barn owl populations or subspecies may have evolved different dispersal patterns in response to differing nest site availabilities or to fluctuations in food supply of differing severity.

Many barn owl populations seem to show highly variable and unpredictable juvenile and adult mortality and the available evidence suggests that most die directly of starvation caused by a decline in the population of their preferred prey, the birds' limited ability to switch to alternative prey and their tendency to remain rather than move off in search of other areas with good food supplies. In northern areas, high mortality also occurs unpredictably in association with exceptionally severe winters and presumably, in hot climates, death rates increase following severe droughts. Average annual mortality rates of barn owls generally exceed those of other similar sized owls and raptors inhabiting the same areas and periodically their mortality is very much greater. At times a breeding barn owl's chances of surviving to the next breeding season are less than its chances of dying. One might therefore expect that natural selection would have favoured those owls that invest more heavily in the production of offspring when the opportunity arises. Any negative effect on their subsequent survival that might be incurred by increasing their reproductive effort would be of reduced importance given that they would have a very high expectation of death from starvation in any case.

Compared with other owls and raptors that are less specialist in their diet and that can, by broadening their diet, maintain higher survival rates through periods of food shortage, barn owls do indeed show many differences that enable them to achieve a higher reproductive output. They mature quickly: the earliest bird in the Scottish study laid her first clutch only 4 months after fledging and most of those that succeeded in breeding at all produced their first eggs in the season immediately following the one in which they were born. Delayed breeding was extremely rare and once established as breeders, mated pairs very seldom missed out a breeding season; non-breeding occurred only under conditions of very extreme food shortage.

If birds are to respond quickly by breeding whenever food supplies and weather conditions permit, it would be advantageous for them to have food supply acting directly on them as a proximate stimulus to breeding rather than to have their breeding controlled by day length changes or some other strictly regular or seasonally predictable stimulus. Barn owls in tropical areas may lay at any time of year and in temperate areas laying dates vary widely from year-to-year in relation to food supply. In Europe, for example, laying may begin as early as February in peak vole years but may not start until June in vole decline years. Within years, there is also an exceptionally wide spread of laying. This all suggests that breeding in barn owls may indeed be controlled directly by food supply rather than by changes in day length. Their response, if any, to photoperiod has not been tested but this could easily be done.

Barn owl eggs are considerably smaller than expected on the basis of female weight. Using the same reserves or daily levels of food intake this should enable them to produce more eggs. This could increase clutch size when food conditions are good but perhaps just as importantly might allow the production of at least some eggs when food supplies are low. The eggs hatch asynchronously, thus enabling a rapid response by brood reduction rather than the loss of the entire brood should food conditions deteriorate during the breeding attempt. The number of young produced can thus be kept high. Should food availability remain high towards the end of rearing the first brood, a second and sometimes even a third clutch of eggs may be laid. In Europe, some pairs may be involved in some aspect or other of breeding for up to 8 months of the year and may produce as many as 10 or 12 young, much more than any other similar sized bird of prey in the same areas.

The pattern of moult of barn owls living in temperate regions is unusual, involving the replacement of only one primary and a small number of secondaries and tail feathers at the first moult, which takes place during their first breeding season. Older birds normally moult more feathers, but never all in one season. Has this pattern evolved to reduce the energetic demands of moult and minimise possible conflict with the demands of breeding? First year birds, being less experienced, are less likely than older birds to be able to meet high demands from both and also, as survival between their first and second years is less than between subsequent years, natural selection may have favoured those that are able to invest more in reproduction in their first year. There must be a disadvantage, or cost, associated with not replacing flight feathers, otherwise they would never be replaced, but the losses arising from this are presumably more than offset from any gains perhaps in the production of more young.

Much of the available evidence suggests that, in association with high mortality rates resulting from their relatively inflexible predation behaviour and their specialist cavity nesting requirements which probably discourages movements, barn owls have evolved a life history pattern that involves a much higher investment in reproduction than is found in many other medium sized owls and raptors. However, the general applicability of these ideas to all barn owl subspecies cannot be ascertained at present. Those that have evolved in more stable tropical environments may be rather different. For these subspecies there is some published information on breeding but nothing on survival rates or movements and this would be a very fruitful area for further research.

Barn owl conservation

In temperate areas, barn owls are often thought of as vulnerable birds, because of their periodically very high mortality rates and sometimes dramatic declines in numbers, but they are actually amazingly resilient. Their very high reproductive potential enables them to increase rapidly when prey abundance is high. In a conservation sense, their vulnerability does not lie in the details of their population dynamics or in their life history patterns, indeed those are their strength, but rather in their highly specialised predation behaviour and habitat requirements. They are extremely sensitive to changes in agricultural landscapes that reduce the distribution and abundance of their preferred small mammal prey. The conservation of barn owls is not particularly complex; if the appropriate types of grassland foraging habitats containing populations of small mammal prey are provided in appropriate quantities and spatial distributions, and if suitable nest sites are available, the owls have the natural ability to increase quickly and to maintain themselves.

In a country such as Britain where the distribution of barn owls is now patchy and overall numbers substantially diminished, a widespread recovery will be achieved only through extensive improvements in the quantity and quality of habitat available to the owls. However, results could be attained more rapidly and more effectively by the application of an improved understanding of the owls' ecology and by more co-ordinated planning. Cooperation among neighbouring farmers and foresters is needed so that adequate habitat improvements are carried out throughout the whole of specifically targeted areas that are large enough to support viable populations of the owls. The most sensible places to start are those that are already known to have good-sized populations and the aim in these areas should be to improve the habitat beyond that needed simply to achieve population stability so that the production of young considerably exceeds the level required to replace local losses. Only in this way will a surplus be available to produce a gradual extension and infilling of the population and the restocking of adjacent depleted areas. Probably an average productivity of around 4 to 5 fledged young per pair should be aimed for in these priority conser-vation areas and, to attain the scale of habitat change needed for this, significant financial incentives may have to be offered to a high pro-portion of the farmers. There is a very real possibility that unless resources are focussed intensively, at least initially, on a number of specific areas in this way, the effort will be spread too thinly to achieve any long lasting results.

Finally, a small word of caution. Many plants and animals in farmland areas are under threat and there is a risk that because barn owls are attractive and appealing, they may be conserved at the expense of less obvious species. Resources for conservation will always be limited and it is important that a holistic view be taken so that as many as possible are cared for. Fully integrated strategies are needed for the conservation of all wildlife in farmland areas, and for farming as well for, without the sustained efforts and commitment of those who live and work in the countryside, little will be achieved.

Appendix 1. *Weights and measurements for 22 barn owl subspecies. Number of observations in parentheses*

Subspecies	Body mass (g)		Wing length (mm)		Tarsus length (mm)		Tail length (mm)		Reference
	Male	Female	Male	Female	Male	Female	Male	Female	
T.a. alba	330 (361)	370 (445)	293 (139)	293 (110)	57 (139)	58 (110)	115 (139)	115 (110)	Taylor, unpublished
T.a. guttata	306 (60)	357 (122)	286 (54)	289 (66)	57 (49)	56 (61)	115 (23)	114 (22)	Cramp 1985
T.a. affinis	329 (11)	363 (7)	293 (11)	294 (7)	63 (11)	64 (7)	114 (11)	116 (7)	Fry et al. 1988
T.a. pratincola	474 (112)	566 (166)	327 (103)	328 (155)	71 (85)	72 (80)	139 (85)	141 (81)	Marti 1990
T.a. delicatula	298 (14)	328 (28)	279 (15)	281 (30)	–	70 (8)	117 (15)	118 (31)	Boles, unpublished
T.a. javanica	555 (13)	612 (26)	300 (13)	300 (26)	70 (1)	–	134 (1)	–	Wells, unpublished
T.a. contempta	–			280 (17)		67 (9)		127 (13)	de Groot 1983
T.a. tuidara	–			282 (11)		66 (9)		131 (6)	de Groot 1983
T.a. punctatissima	–	264 (2)	224 (16)	223 (28)	54 (16)	55 (23)	120 (16)	121 (23)	de Groot 1983
T.a. ernestii	–			293 (8)		62 (8)		118 (8)	Cramp 1985

Taxon		Wing		Tail		Bill		Reference
T.a. erlangeri	–	294 (23)		62 (11)		114 (23)		Cramp 1985
T.a. detorta	–	290 (20)		–		113 (16)		Cramp 1985
T.a. gracilirostris	–	257 (4)		–		99 (4)		Cramp 1985
T.a. schmitzi	–	280 (9)		–		112 (9)		Cramp 1985
T.a. guatemalae	–	309 (10)	317 (10)	74 (10)	73 (10)	127 (10)	130 (10)	Wetmore 1968
T.a. stertens	–	262–303 (10)	278–313 (5)	70 (?)		111–23 (?)	114–25 (?)	Ali and Ripley 1969
T.a. thomensis	–	254 (13)		64 (6)	64 (6)	102 (4)	107 (5)	de Naurois 1983
T.a. deroepstorffi	–	260 (?)		61 (?)		113 (?)		Ali and Ripley 1969
T.a. lulu	–	272 (?)		64 (?)		112 (?)		Pont 1975
T.a. furcata	–	342 (9)	353 (4)	–		134 (12)	138 (7)	Parkes and Phillips 1978
T.a. niveicauda	–	331 (5)	347 (1)	–		126 (6)	128 (1)	Parkes and Phillips 1978
T.a. bondi	–	301 (1)	316 (1)	–		114 (1)	114 (1)	Parkes and Phillips 1978

Appendix 2. *Analysis of 52 diet samples from barn owl populations around the world to show the relative importances of small mammal and rodent prey taken, the most important species in each area and the range of prey weights taken*

Country	Latitude	Diet content (% by number)		Total number terrestrial small mammal species	Most frequent prey species (by number)		Most important prey species (by biomass)		No. of species forming 80% of diet by		Prey weight (g)		Source
		Small mammals	Rodents		Species	% of diet	Species	% of diet	Number	Biomass	Normal range	Average	
Europe													
Scotland	55°N	99.3	80.0	8	Microtus agrestis	44.7	M. agrestis	64.9	3	2	4–60	20	Taylor, this study
England	53°N	97.6	89.7	11	M. agrestis	50.5	M. agrestis	63.3	3	3	4–60	21	Glue 1974
Wales	53°N	90.0	54.0	7	M. agrestis	42.0	M. agrestis	51.6	3	6	4–25	20	Brown 1981
Ireland	52°N	96.4	67.7	11	Apodemus sylvaticus	55.6	Apodemus sylvaticus	50.3	3	3	4–60	22	Smal 1987
Denmark	55°N	94.2	55.8	15	Sorex araneus	34.4	M. agrestis	24.0	5	5	4–60	17	Lange 1948
Germany	53°N	99.5	80.1	11	M. arvalis	76.1	M. arvalis	85.2	2	1	4–30	18	Bohnsack 1966
	51°N	97.7	46.7	13	M. arvalis	33.6	M. arvalis	42.9	5	5	4–30	15	Knorre 1973
	49°N	98.0	57.4	15	M. arvalis	25.0	M. arvalis	32.1	4	4	4–40	18	Bethge and Hayo 1979
Holland	52°N	97.6	44.3	16	S. araneus	39.4	M. arvalis	41.6	4	4	5–40	14	de Bruijn 1979
Belgium	51°N	96.2	39.6	10	S. araneus	36.8	S. araneus	25.8	5	4	4–30	13	Straeten and Asselberg 1973
France	48°N	99.0	71.0	18	M. arvalis	47.9	M. arvalis	56.5	3	3	4–40	20	Cuisin and Cuisin 1979
	47°N	98.0	64.3	18	M. arvalis	39.5	M. arvalis	49.0	4	4	4–40	19	Baudvin 1983
	43°N	81.6	48.0	15	Crocidura russula	31.4	Apodemus spp.	25.8	6	5	7–80	16	Libois et al. 1983
Switzerland	46°N	99.7	66.5	14	M. arvalis	30.8	M. arvalis	34.6	6	5	4–70	20	Pricam and Zelenka 1964
Poland	52°N	95.0	76.1	12	Mus musculus	41.0	Mus musculus	45.7	4	4	4–30	17	Ruprecht 1964
Czechoslovakia	50°N	89.4	88.7	6	M. arvalis	84.2	M. arvalis	89.0	1	1	4–50	25	Pikula et al. 1984
Hungary	47°N	94.0	81.3	17	M. arvalis	50.2	M. arvalis	52.2	4	3	4–30	20	Schmidt 1973
Poland	53°N	85.0	67.6	17	Mus musculus	26.4	Mus musculus	32.5	5	4	7–40	16	Ruprecht 1979b
Ukraine	50°N	96.0	82.9	17	Mus musculus	34.2	Mus musculus	34.7	4	4	4–30	19	Pidoplitschka 1937

Spain	42°N	96.8	79.2	10	*Mus* spp.	44.4	*Mus* spp.	37.3	4	4	4–30	17	Brunet-Lecompte and Delibes 1984
Italy	38°N	87.1	68.8	12	*Mus musculus*	44.9	*Mus musculus*	37.0	4	4	10–30	20	Herrera 1974
	43°N	96.1	77.4	12	*Apodemus* spp.	33.7	*Apodemus* spp.	49.0	5	3	5–80	18	Lovari et al. 1976
Palestine	32°N	77.7	65.0	11	*M. guntheri*	46.1	*M. guntheri*	–	3	–	–	–	Dor 1947
Canary Is	28°N	74.5	74.5	2	*Mus musculus*	70.3	*Mus musculus*	76.9	2	2	1–60	17	Martin et al. 1985
Morocco	32°N	99.7	88.7	4	*Mus musculus*	69.7	*Mus musculus*	–	4	–	–	–	Saint-Girons and Thouy 1978
North and South America													
British Columbia	48°N	97.0	80.1	18	*M. townsendii*	73.0	*M. townsendii*	82.5	1	1	6–65	55	Campbell et al. 1987
Washington	46°N	97.8	97.8	9	*Perognathus parvus*	63.9	*Thomomys talpoides*	39.3	3	3	10–100	28	Knight and Jackman 1984
Oregon	45°N	99.7	57.2	10	*S. vagrans*	42.3	*M. townsendii*	71.6	2	2	6–60	29	Giger 1965
Idaho	43°N	98.5	93.5	15	*M. montanus*	55.3	*M. montanus*	55.7	3	3	10–70	34	Marti 1988
Utah	41°N	97.0	92.6	18	*Microtus* spp. (*montanus* and *pennsylvanicus*)	72.0	*Microtus* spp.	84.4	2	2	10–50	38	Marti 1988
Michigan	43°N	98.9	92.0	13	*M. pennsylvanicus*	85.0	*M. pennsylvanicus*	93.0	1	1	15–40	38	Wallace 1948
Ohio	41°N	98.5	74.5	17	*M. pennsylvanicus*	63.9	*M. pennsylvanicus*	75.8	2	2	18–90	33	Colvin and McLean 1986
New Jersey	39°N	98.7	89.0	11	*M. pennsylvanicus*	74.1	*M. pennsylvanicus*	80.4	1	2	15–150	35	Colvin 1984
Virginia	38°N	94.0	76.0	11	*M. pennsylvanicus*	55.0	*M. pennsylvanicus*	60.0	3	3	5–65	42	Rosenburg 1986
Colorado	40°N	98.6	97.0	13	*M. ochrogaster*	34.8	*M. ochrogaster*	30.1	6	5	8–130	46	Marti 1974
Oklahoma	36°N	100	100	9	*Perognathus* spp.	42.0	*Perognathus* spp.	29.2	4	4	18–120	55	Ault 1971
California	37°N	98.5	98.5	7	*Perognathus irroratus*	44.6	*Thomomys bottae*	71.4	2	2	20–100	–	Fitch 1947
Georgia, Alabama, S. Carolina	33°N	96.5	78.9	14	*Sigmodon hispidus*	54.8	*Sigmodon hispidus*	–	1	1	20–115	76 (?)	French and Wharton 1975
Texas	29°N	92.0	82.5	13	*Sigmodon hispidus*	34.2	*Sigmodon hispidus*	69.8	3	7	2–115	58	Byrd 1982
	28°N	85.4	66.6	10	*Baiomys taylori*	19.7	*Sigmodon hispidus*	28.0	6	8	5–120	65	Otteni et al. 1972

Appendix 2 (*cont.*)

Country	Latitude	Diet content (% by number)		Total number terrestrial small mammal species	Most frequent prey species (by number)		Most important prey species (by biomass)		No. of species forming 80% of diet by		Prey weight (g)		Source
		Small mammals	Rodents		Species	% of diet	Species	% of diet	Number	Biomass	Normal range	Average	
Chile		95.1	90.1	9	*Oryzomys longicaudatus*	28	*Oryzomys longicaudatus*	?	5	?	17–70	60	Herrera and Jaksic 1980
Africa													
Transvaal		92.9	85.8	25	*Praomys natalensis*	45.9	*Praomys natalensis*	45.8	6	2	7–110	42	Vernon 1972
Cape Province		91.6	85.6	20	*Gerbillus paeba*	28.3	*Desmodillus auricularis*	25.1	6	10	10–110	39	Vernon 1972
	30°S	96.8	90.0	13	*Praomys natalensis*	21.0	*Otomys irroratus*	29.0	6	5	7–110	63	Perrins 1982
Natal		96.9	81.2	12	*Praomys natalensis*	58.0	*Praomys natalensis*	60.4	5	2	7–110	42	Vernon 1972
Namibia	25°S	89.5	85.2	11	*Gerbillus* and *Tatera* spp.	74.9	*Gerbillus* and *Tatera* spp.	63.0	3	6	7–50	27	Vernon 1972
Australia													
Victoria		99.1	97.8	4	*Mus musculus*	96.3	*Mus musculus*	80	1	1	15–80	20	Morton 1975
NSW and S. Australia	28°S	98.0	94.9	10	*Mus musculus*	82.8	*Mus musculus*	64.5	1	2	15–100	19	Morton and Martin 1979
S.W. Queensland	25°S	99.3	94.8	10	*Rattus villosissimus*	45.1	*Rattus villosissimus*	75.6	4	2	20–100	60	Morton *et al.* 1977
Northern Territory	20°S	85.5	80.3	10	*Notomys alexis*	41.1	*Notomys alexis*	54.5	7	5	10–30	23	Smith and Cole 1989
Other areas													
Malaysia	2°N	99.2	98.2	8	*Rattus tiomanicus*	89.8	*Rattus tiomanicus*	90	1	1	–	80	Lenton 1984a
Galapagos Islands	1°S	44.9	44.9	2	*Mus musculus*	39.3	*Mus musculus*	47.6	2	2	2–60	25	de Groot 1982

Appendix 3. *A number of sample barn owl diets from Europe and North America to show precise details of species composition*

A. European countries

Prey items eaten (% of total no. eaten)

Species	Canary Is.	Czechoslovakia	Poland	Spain	Italy	Belgium	Holland	Germany	Germany	France (Mediterranean)	France (Burgundy)	Ireland	Wales	England	Scotland
Talpa europaea			+			+	+	+	+	+	+		+	+	+
Sorex araneus		+	14.1		1.8	36.8	36.6	18.6	18.6		18.1		29.4	23.7	38.3
Sorex minutus			1.4		4.4	1.8	1.4	1.3	1.3		2.0	11.9	5.4	4.7	1.4
Neomys fodiens			1.5				+	+	+		+			1.3	+
Neomys anomalus				+											
Crocidura leucodon					4.6										
Crocidura russula				14.9	4.8	14.5	10.0	20.2		31.4	12.5				
Crocidura sauveolens				2.4	3.0					1.3					
Suncus etruscus				+	+					+					
Chiroptera spp.		+	+							+				+	
Eliomys quercinus				+						+	+				
Glis glis				+						+	+				
Muscardinus avellanarius					2.2					+	+				
Clethrionomys glareolus			+		10.2	1.2	+	1.6			3.0	13.7	4.2	3.3	2.1
Microtus oeconomus			5.6				+								
Microtus agrestis			+			11.9	2.9	16.3	1.7	4.8	4.8		47.8	50.7	44.7
Microtus arvalis		84.2	23.1			12.4	26.8	25.0	76.1		39.5				
Arvicola terrestris							+	2.2	+		+		+	+	
Arvicola sapidus				+						+	+				
Pitymys subterraneus		+				4.3	+				+				
Pitymys				11.9						2.0					
Pitymys duodecimcostatus										+					
Pitymys savii					26.8										
Micromys minutus			3.5				1.6	+		+	+		+	3.2	
Apodemus spp.		1.4	4.3	10.8	33.7	4.8	5.0	11.3	1.0	20.7	14.2	55.6	8.3	6.2	12.6
Rattus rattus			+	+	+					+					
Rattus norvegicus			+	+			+	+					1.0		
Rattus spp.	4.1	1.1							1.8			9.0		3.2	+
Mus spp.	70.3	1.5	26.4	45.0	4.1	2.8	2.3	+		17.3	+	5.8	+	+	+
Mustela nivalis						+	+	+	+	+	+		+	+	+
Aves	2.2	+	13.2	4.3	2.0	2.9	2.0	+	+	10.5	+	2.2	1.1	2.3	+
Reptilia	2.5		+	+											
Amphibia	+		1.9	4.3			+		+	6.8	1.1	+		+	+
Insecta	18.3		+	4.3	2.0		+		1.0	1.0	+	+			
Total sample size (No. of items)	2058	913	16944	14801	3661	34866	92383	33469	11913	18561	40614	8229	4575	16350	40294
Reference	Martin *et al.* 1985	Pikula *et al.* 1984	Ruprecht 1964	Herrera 1974	Lovari *et al.* 1976	Straeten and Asselberg 1973	De Bruijn 1979	Bethge and Hayo 1979	Bohnsack 1966	Libois *et al.* 1983	Baudvin 1983	Smal 1987	Glue 1974	Buckley and Goldsmith 1975	Taylor this study

Appendix 3 (*cont.*)
B. North American areas

	Prey items eaten (% of total no. eaten)								
	Texas	Oklahoma	Virginia	New Jersey	Ohio	Colorado	Utah	Idaho	British Columbia, Canada
Species									
Shrews, *Sorex* spp.				+	+		4.7	+	11.7
Masked shrew *S. cinereus*									1.5
Vagrant shrew *S. vagrans*									3.1
Short-tailed shrew *Blarina brevicauda*	16.3		13.0	7.6	20.2				
Least shrew *Cryptotis parva*		+	6.0		+	+			1.0
Moles Talpidae				+	+				
Bats Chiroptera	2.7		+	+	+		+	+	+
Cotton tails *Sylvilagus* spp.					+	1.5			
Eastern chipmunk *Tamias striatus*									
Pocket gophers *Thomomys* spp.	4.0	1.1				7.3	+	5.6	
Pocket mice *Perognathus* spp.	9.6	42.1				7.3	+	6.6	
Kangaroo rat *Dipodomys* spp.						4.0		8.2	
Rice rat *Oryzomys* spp.	5.8		1.0	4.2					
Harvest mice *Reithrodontomys* spp.	10.2	18.5	2.0			7.3	6.4	3.5	
White-footed mice *Peromyscus* spp.	1.8	14.1	1.0	2.0	5.0	25.1	7.7	12.8	1.7
Grasshopper mice *Onychomys* spp.		8.4				+		+	
Cotton rat *Sigmodon* spp.	10.8	11.8	+						
Wood rat *Neotoma* spp.	4.7	3.9				+		+	
Meadow vole *Microtus pennsylvanicus*			55.0	74.1	63.9	9.5			
Mountain vole *M. montanus*									73.0
Townsend's vole *M. townsendii*									1.3
Oregon vole *M. oregoni*									
Prairie vole *M. ochrogaster*						34.8	72.0		
Voles *Microtus* spp.				+	+	+	+	55.3	
Muskrat *Ondatra*				+			+	+	
Ship rat *Rattus rattus*							+		1.0
Norway rat *R. norvegicus*			2.0	1.9	+				1.3
House mouse *Mus musculus*			12.0	6.7	+	+	5.3	6.9	1.1
Jumping mouse *Zapus hudsonius*				1.1	4.1				
Pygmy mouse *Baiomys tailori*	19.7								
Aves	12.9		6.0	1.3	1.5	1.4	2.5	1.5	1.9
Reptilia									
Amphibia	1.7								
Insecta							+		
Number of items	11408	680	1061	9531	12589	4366	25330	38199	30218
Reference	Otteni *et al.* 1972	Ault 1971	Rosenburg 1986	Colvin 1984	Colvin and McLean 1986	Marti 1974	Marti 1988	Marti 1988	Campbell *et al.* 1987

+ denotes values for the prey eaten of less than 1% of the total number.

References

Alegre, J. Hernandez, A., Purroy, F.J. and Sanchez, A.J. (1989). Distribucion altitudinal y patrones de afinidad trofica geografica de la lechuza comun (*Tyto alba*). en Leon. *Ardeola*, **36**, 41–54.

Ali, S. and Ripley, S.D. (1969). *Handbook of the birds of India and Pakistan*. Oxford University Press, Bombay.

Anonymous (1984). Status reports: state endangered and threatened raptor species. *Eyas*, **7**, 17–20.

Arbib, R. (1979). The blue list for 1980. *American Birds*, **33**, 830–835.

Ashmole, M.P. (1962). The Black Noddy (*Anous tenuirostris*) on Ascension Island. *Ibis*, **103**, 235–319.

Ault, J.W. (1971). *A quantitative estimate of barn owl nesting habitat quality*. M.Sc. Thesis, Oklahoma State University, Stillwater, Okla.

Bairlien, F. von (1985). Dismigration und Sterblkeit in Suddeutschland beringter Schleiereulen (*Tyto alba*). *Die Vogelwarte*, **33**, 81–108.

Baudvin, H. (1975). Biologie de reproduction de la Chouette effraie (*Tyto alba*). *Le Jean-le-Blanc*, **14**, 1–51.

Baudvin, H. (1978). Les causes d'échec des nichées de chouette effraie (*Tyto alba*). *Le Jean-le-Blanc*, **17**, 29–34.

Baudvin, H. (1980). Les surplus proies au site de nid chez la Chouette effraie (*Tyto alba*). *Nos Oiseaux*, **35**, 232–38.

Baudvin, H. (1983). La régime alimentaire de la Chouette Effraie (*Tyto alba*). *Le Jean-le-Blanc*, **22**, 1–108.

Baudvin, H. (1986). La reproduction de la Chouette Effraie (*Tyto alba*). *Le Jean-le-Blanc*, **25**, 1–125.

Baumler, W. (1975). Activity of some mammals in the field. *Acta Theriologica*, **20**, 365–379.

Bent, A.C. (1938). Life histories of North American birds of prey. 2. *Bulletins of the US National Museum*, **170**, 1–482.

Berger, A.J. (1972). *Hawaiian Birdlife* University Press of Hawaii, Honolulu.

Bersuder, B. and Kayser, Y. (1988). La prédation des chiroptères par la Chouette effraie (*Tyto alba*) en Alsace et dans les contrées Limitrophes. *Ciconia*, **12**, 135–152.

Bethge, E. and Hayo, L. (1979). Untersuchungen an einer population der Schleiereule *Tyto alba* in einem ländlichen Bezirk des westlichen Saarlands. *Anzeiger der Ornithologischen Gesellschaft in Bayern*, **18**, 161–170.

Birkhead, T.R. and Muller, A.P. (1990). *Sperm competition in birds*. Academic Press, London.

Bishop, J.A. and Hartley, D.J. (1976). Ecology of warfarin resistant rats. *Journal of Animal Ecology*, **45**, 623–646.

Blaker, G.B. (1933). The Barn Owl in England. *Bird Notes and News*, **15**, 169–172, 207–211.

Bohnsack, P. (1966). Über die Ernahrung der Schleiereule, *Tyto alba*, insbesondore auselhalb der Brutzeit, in einem westholsteinischen Massenwechselgebiet der Feldmaus, *Microtus arvalis*. *Corax*, **1**, 162–172.

Bonnet, P. (1932). An outlaw barn owl. *Condor*, **30**, 320.

Borquin, J-D. (1983). Mortalité des rapaces de long de l'autoroute Genève-Lausanne. *Nos Oiseaux*, **37**, 149–169.

Boyce, C.C. and Boyce III, J.L. (1988a). Population biology of *Microtus arvalis* I. Lifetime reproductive success of solitary and grouped females. *Journal of Animal Ecology*, **57**, 711–722.

Boyce, C.C. and Boyce III, J.L. (1988b). Population biology of *Microtus arvalis* II. Natal and breeding dispersal of females. *Journal of Animal Ecology*, **57**, 723–736.

Boyce, C.C. and Boyce III, J.L. (1988c). Population biology of *Microtus arvalis* III. Regulation of numbers and breeding dispersion of females. *Journal of Animal Ecology*, **57**, 737–754.

Braaksma, S. (1980). Gegevens over de achteruitgang van de kerkuil (*Tyto alba guttata* – Brehm) in West-Europa. *Wielwaal*, **46**, 421–428.

Braaksma, S. and de Bruijn, O. (1976). De kerkuilstand in Nederland. *Limosa*, **49**, 135–187.

Brodkorb, P. (1972). Catalogue of fossil birds. 4. *Bulletin of the Florida State Museum Biology Society*, **15**, 163–266.

Brown, D.J. (1981). Seasonal variations in the prey of some barn owls in Gwynedd. *Bird Study*, **28**, 139–146.

Brown, L.E. (1956). Field experiments on the activity of the small mammals, *Apodemus*, *Clethrionomys* and *Microtus*. *Proceedings of the Zoological Society of London*, **126**, 549–564.

Brunet-Lecompte, P. and Delibes, M. (1984). Alimentacion de la lechuza comun *Tyto alba* en la cuenca del Duero, Espana. *Donana, Acta Vertebrata* **11**, 213–229.

Buckley, J. and Goldsmith, J.G. (1975). The prey of the Barn Owl (*Tyto alba alba*) in east Norfolk. *Mammal Review*, **5**, 13–16.

Buden, D.W. (1974). Prey remains of barn owls in the southern Bahama Islands. *Wilson Bulletin*, **86**, 336–343.

Buhler, P. (1970). Schlupfhilfe-Verhalten bei der Schleiereule (*Tyto alba*). *Die Vogelwelt*, **91**, 121–130.

Buhler, P. and Epple, W. (1980). The vocalisations of the barn owl. *Journal für Ornithologie*, **121**, 36–71.

Bunn, D.S., Warburton, A.B. and Wilson, R.D.S. (1982). *The Barn Owl*. T & A.D. Poyser, Carlton, UK.

Byrd, C.L. (1982). Home range, habitat and prey utilisation of the barn owl in south Texas. M.Sc. Thesis. Texas A. and I. University, Kingsville, Texas.

Campbell, R.W., Manuwal, D.A. and Harestad, A.S. (1987). Food habits of the common Barn Owl in British Columbia. *Canadian Journal of Zoology*, **65**, 578–586.

Carpenter, M.L. and Fall, M.W. (1967). The barn owl as a red-winged blackbird predator in north west Ohio. *Ohio Journal of Science*, **67**, 317–318.

Carstairs, J.L. (1974). The distribution of *Rattus villosissimus* (Waite) during plague and non-plague years. *Australian Wildlife Research*, **2**, 95–106.

Cavé, A.J. (1968). The breeding of the kestrel, *Falco tinnunculus* L. in the reclaimed area Oostelijk Flevoland. *Netherlands Journal of Zoology*, **18** 313–407.

Chanson, J.M., Bourbet, P., Giraudoux, P., Michaud, G. and Michelat D. (1988). Étude sur la reproduction et les deplacements de la chouette effraie (*Tyto alba*) en Franche-Compté: réflexions méthodologiques. *Alauda*, **56**, 197–225.

Charnov, E.L. (1976). Optimal foraging: attack strategy of a mantid. *American Naturalist*, **110**, 141–151.

Chitty, D. and Chitty, H. (1962). Population trends among voles at Lake Vyrnwy. *Symposium Theriologica Brno* (1960), pp. 67–76.

Churchfield, S. (1981). Water and fat contents of British shrews and their role in the seasonal changes in body weight. *Journal of Zoology* (London), **194**, 165–173.

Churchfield, S. (1982). The influence of temperature on the activity and food consumption of the common shrew. *Acta Theriologica*, **27**, 295–304.

Churchfield, S. (1984). An investigation of the population ecology of syntopic shrews inhabiting watercress beds. *Journal of Zoology* (London), **204**, 229–240.

Churchfield, S. (1990). *The Natural History of Shrews*. Christopher Helm, London.

Clarke, J.R. (1956). The aggressive behaviour of the vole. Behaviour, **9**, 1–23.

Clarke, R. (1975). A field study of the short-eared owl (*Asia flammeus*. Pontop.) in North America. *Wildlife Monographs*, **47**, 1–67.

Classens, A.J.M. and O'Gorman, F. (1965). The Bank Vole (*Clethrionomys glareolus* Schreber): a mammal new to Ireland. *Nature* (London), **205**, 923–924.

Cody, C.B.J. (1982). Studies on behavioural and territorial factors relating to the dynamics of woodland rodent populations. D.Phil. Thesis, University of Oxford.

Collinge, W.E. (1924). The food of some British wild birds. *A study of Economic Ornithology*, 2nd edn. York, privately printed.

Colvin, B.A. (1984). Barn owl foraging behaviour and secondary poisoning hazard from rodenticide use on farms. Ph.D. Thesis, Bowling Green State University.

Colvin, B.A. (1985). Common barn owl population decline in Ohio and the relationship to agricultural trends. *Journal of Field Ornithology*, **56**, 224–235.

Colvin, B.A. and Hegdal, P.L. (1986–90). *Annual summaries of barn owl field activities*. Denver Wildlife Research Centre, Denver, Colo.

Colvin, B.A. and McLean, E.B. (1986). Food habits and prey specificity of the common barn owl in Ohio. *Ohio Journal of Science*, **86**, 76–80.

Cooke, A.S., Bell, A.A. and Haas, M.B. (1982). *Predatory birds, pesticides and pollution*. Institute of Terrestrial Ecology, Cambridge.

Cramp, S. (ed.) (1985). *Handbook of the Birds of Europe, the Middle East and North Africa*, vol 4. Oxford University Press, Oxford.

Cuisin, J. and Cuisin, M. (1979). Le régime alimentaire de la Chouette effraie (*Tyto alba*) dans le canton de Riceys (Auge) et ses environs immédiats. *Orfo*, **49**, 81–89.

Daan, S. and Slopsema, S. (1978). Short-term rhythms in foraging behaviour of the common vole, *Microtus arvalis*. *Journal of Comparative Physiology*, **127**, 215–227.

Davis, D.H.S. (1933). Rhythmic activity in the short-tailed vole, *Microtus agrestis*. *Journal of Animal Ecology*, **2**, 232–238.

Davis, D.H.S. (1959). The barn owl's contribution to ecology and palaeoecology. *Ostrich* (suppl 3), 144–153.

Dawe, N.K., Runyan, C.S. and McKelvey, R. (1978). Seasonal food habits of the barn owl (Tyto alba) on the Alaskan National Wildlife Area, British Columbia. *Canadian Field Naturalist*, **92**, 151–155.

Dean, W.R.J. (1972). Age distribution of *Praomys natalensis* prey in *Tyto alba* pellets. *Zoologica Africana*, **8**, 140.

de Bruijn, O. (1979). Feeding ecology of the barn owl *Tyto alba* in the Netherlands. *Limosa*, **52**, 91–154.

de Groot, R.S. (1983). Origin, status and ecology of the owls in Galapagos. *Ardea*, **71**, 167–182.

de Jong, J. (1983). *De Kerkuil*. Kosmos, Utrecht.

de Jong, J. (1991). Protection et recherches sur la consommation alimentaire et le bilan énergétique chez la Chouette effraie, *Tyto alba*. In *Rapuces Nocturnes*, pp. 109–122. Actes du 30 Colloque interrégional d'ornithologie Porrentruy (Suisse).

de Naurois, R. (1982). Le statut de la Chouette effraie de l'archipel du Cape Verte. *Rivista Italiana di Ornitologia*, **52**, 154–166.

de Naurois, R. (1983). Falconidae, Psiltacidae, et Strigiformes des îles de Sao Thome' et Principe. (Golfe de Guinée). *Bonner Zoologische Beitraege*, **34**, 429–446.

de Wavrin, H. (1977). Diminution des effectifs de Chouette Effraie (*Tyto alba*) en moyenne Belgique. *L'Homme et l'Oiseaa*, **15**, 19–22.

Derting, T.L. and Cranford, J.A. (1989). Physical and behavioural correlates of prey vulnerability to barn owl (*Tyto alba*) predation. *American Midland Naturalist*, **121**, 11–20.

Dickman, C.R., Predavec, M. and Lynam, A.J. (1991). Differential predation of size and sex classes of mice by the barn owl *Tyto alba*. *Oikos*, **62**, 67–76.

Dijkstra, C., Vuuersteen, L., Daan, S. and Masman, D. (1982). Clutch size and laying date in the kestrel *Falco tinnunculus*: effect of supplementary food. *Ibis*, **124**, 210–213.

Dobson, P. and Wexlar, D. (1979). Taphonomic investigations of owl pellets. *Paleobiology*, **5**, 275–284.

Dor, M. (1947). Observations sur les micromammifières trouvés dans les pelotes de la Chouette effraie *Tyto alba* en Palestine. *Mammalia*, **11**, 50–54.

Doucet, J.G. and Bider, J.R. (1969). Activity of *Microtus pennsylvanicus* related to moon phase and moonlight revealed by the sand transect technique. *Canadian Journal of Zoology*, **47**, 1183–1186.

Doucet, J.G. and Bider, J.R. (1974). The effects of weather on the activity of the masked shrew. *Journal of Mammalogy*, **55**, 348–363.

Drent, R.H. and Daan, S. (1980). The prudent parent: energetic adjustments in avian breeding. *Ardea*, **68**, 225–252.

Duckett, J.E. (1976). Owls as major predators of rats in oil palm estates with particular reference to the barn owl (*Tyto alba*). *Planter* (Kuala Lumpur), **52**, 4–15.

Duckett, J.E. (1984). Barn Owls (*Tyto alba*) and the 'second generation' rat-baits utilised in oil palm plantations in Peninsular Malaysia. *Planter* (Kuala Lumpur), **60**, 3–11.

Duckett, J.E. (1991). Management of the barn owl (*Tyto alba javanica*) as a predator of rats in oil palm (*Elaeis quineensis*) plantations in Malaysia. *Birds of Prey Bulletin*, **4**, 11–23.

du Pont, J.E. (1975). *South Pacific Birds*. Delaware Museum of Natural History, Delaware.

Edwards, T.C. (1987). Standard rate of metabolism in the common barn owl (*Tyto alba*). *Wilson Bulletin*, **99**, 704–706.

Elton, C. (1927). *Animal Ecology*. Sidgwick and Jackson, London.

Elton, C. (1942). *Voles, Mice and Lemmings*. Oxford University Press, Oxford.

Erkinaro, E. (1969). Der Phasenwachsel der lokomotorischen Aktivitat bei *Microtus agrestis* (L.), *M. arvalis* (Pall.) and *M. oeconomus* (Pall.). *Aquilo* (Series Zoologica), **8**, 1–29.

Erlinge, S., Goransson, G., Hogstedt, G., Jansson, G., Liberg, O., Loman, J., Nilsson, I.N., von Schantz, T. and Sylven, M. (1983). Can vertebrate predators regulate their prey. *American Naturalist*, **123**, 125–133.

Errington, P.L. (1943). An analysis of mink predation upon muskrats in north-central United States. *Iowa Agricultural Research Bulletin*, **320**, 798–924.

Evans, F.C. (1949). A population study of house mice (*Mus musculus*) following a period of local abundance. *Journal of Mammalogy*, **30**, 351–363.

Evans, D.M. (1973). Seasonal variations in the body composition and nutrition of the vole *Microtus agrestis*. *Journal of Animal Ecology*, **42**, 1–18.

Fairley, J.S. (1966). An indication of the food of the short-eared owl in Ireland. *British Birds*, **59**, 307–308.

Fajardo, I. (1990). Mortalidad de la Lechuza comun (*Tyto alba*) en Espana central. *Ardeola*, **37**, 101–106.

Fast, S.J. and Ambrose, H.W. (1976). Prey preference and hunting habitat selection in the barn owl. *American Midland Naturalist*, **96**, 503–507.

Fitch, H.S. (1947). Predation by owls in the Sierran foothills of California. *Condor*, **49**, 137–151.

Fitzgerald, B.M. (1977). Weasel predation on a cyclic population of the montane vole (*Microtus montanus*) in California. *Journal of Animal Ecology*, **46**, 367–397.

Fleay, D. (1968). *Nightwatchmen of Bush and Plain*. Jacaranda Press, Brisbane.

Flowerdew, J.R. (1985). The population dynamics of woodmice and yellow-necked mice. *Symposia of the Zoological Society of London*, **55**, 315–338.

Frank, F. (1953). Zur enstehung übernormaler populationsdichten im Massenwechsel der Feldmaus, *Microtus arvalis* (Pallas). *Zoologisches Jahrbuecher Abteilung fur Systematik Oekologie und Geographie der Tiere*, **81**, 610–624.

Frank, F. (1954). Beiträge zur Biologie der Feldmaus *Mictotus arvalis* (Pallas). Teil 1: Gehegeversuche. *Zoologisches Jahrbuecher Abteilung Systematik*, **82**, 354–404.

Frank, F. (1956). Grundlagen, Möglichkeiten und Methoden der Sanierung von Feldmausplagegebeiten. *Nachrichtenblatt Fuer den Deutschen Pflanzenschutzdienst (Braunschweig)*, **8**, 147–158.

Frank, F. (1957). The causality of microtine cycles in Germany. *Journal of Wildlife Management*, **21**, 113–121.

French, T.W. and Wharton, C.H. (1975). Barn Owls as mammal collectors in Georgia, Alabama and South Carolina. *Oriole*, **40**, 6–10.

Fry, C.H., Keith, S. and Urban, E.K. eds. (1988). *The Birds of Africa* vol 3. Academic Press, London.

Frylestam, B. (1972). Movements and mortality of banded barn owls *Tyto alba* in Scandinavia. *Ornis Scandinavica*, **3**, 45–54.

Fuiczynski, D. (1978), Zur populationsökologie des Baumfalken (*Falco subbuteo*, L. 1758). *Zoologisches Jahrbuecher Systematik Band*, **105**, 193–257.

Genoud, M. and Hausser, J. (1979). Ecologie d'une population de *Crocidura russula* en milieu rural montagnard (Insectivora, Soricidae). *Terre Vie*, **33**, 539–554.

Gessaman, J.A. (1972). Bioenergetics of the snowy owl (*Nyctea scandiaca*). *Arctic and Alpine Research*, **4**, 223–238.

Giger, R.D. (1965). Surface activity of moles as indicated by remains in barn owl pellets. *Murrelet*, **46**, 32–36.

Glue, D. (1973). Seasonal mortality in four small birds of prey. *Ornis Scandinavica*, **4** 97–102.

Glue, D. (1974). Food of the barn owl in Britain and Ireland. *Bird Study*, **21**, 200–210.

Glue, D.E. (1977). Feeding ecology of the Short-eared Owl in Britain and Ireland. *Bird Study*, **23**, 70–78.

Glutz von Blotzheim, U.N. and Schwarzenbach, F.H. (1979). Zur dismigration junger Schleiereule (*Tyto alba*). *Ornithologische Beobachter*, **76**, 1–7.

Goertz, J.W. (1964). The influence of habitat quality upon the density of cotton rat populations. *Ecological Monographs*, **34**, 359–381.

Goodman, S.M. (1986). The prey of barn owls inhabiting the ancient temple complex of Karnak, Egypt. *Ostrich*, **57**, 109–112.

Goszcynski, J. (1976). Estimation of daily food ratio of *Tyto alba*, Scopoli, under natural conditions. *Polish Ecological Studies*, **2**, 95–102.

Goszcynski, J. (1977). Connections between predatory birds and mammals and their prey. *Acta Theriologica*, **22**, 399–430.

Goszcynski, J. (1981). Comparative analysis of food of owls in agrocenoses. *Ekologia Polska*, **29**, 413–439.

Graham, R.R. (1934). The silent flight of owls. *Journal of the Royal Aeronautical Society*, **38**, 837–843.

Greenwood, P.J. (1987). Inbreeding, philopatry and optimal outbreeding in birds. In *Avian Genetics. A Population and Ecological Approach*, eds. F. Cooke and P.A. Buckley, pp. 207–222. Academic Press, London.

Güttinger, H.R. (1965). Gur Winterstblichkeit schweizerischer Schleiereulen (*Tyto alba*) mit besonder Berucksichtingung des Winters 1962/63. *Ornithologische Beobachter*, **62**, 14–23.

Haldane, J.B.S. (1955). The calculation of mortality rates from ringing data. *Proceedings of the International Ornithological Congress*, **11**, 454–458.

Halle, S. (1988). Avian predation upon a mixed community of Common Voles (*Microtus arvalis*) and Wood Mice (*Apodemus sylvaticus*). *Oecologia* (Berlin), **75**, 451–455.

Hamilton, K.L. (1985). Food and energy requirements of captive barn owls *Tyto alba*. *Comparative Biochemistry and Physiology*, **80**, 355–358.

Hansson, L. (1971a). Small rodent food, feeding and population dynamics. A comparison between granivorous and herbivorous species in Scandinavia. *Oikos*, **22**, 183–198.

Hansson, L. (1971b). Habitat, food and population dynamics of the field vole in south Sweden. *Viltrevy*, **8**, 267–378.

Hansson, L. (1977). Spatial dynamics of field voles *Microtus agrestis* in heterogeneous landscapes. *Oikos*, **29**, 539–544.

Hansson, L. (1979). Food as a limiting factor for small rodent numbers: tests of two hypotheses. *Oecologia*, **37**, 297–314.

Hansson, L. (1984). Predation as a factor causing extended low densities in microtine cycles. *Oikos*, **43**, 255–256.

Haukioja, E. and Haukioja, M. (1971). Assessment of the Goshawk (*Accipiter gentilis*) population and its influence in Finland. *Soumen Rusta* **23**, 17–22.

Hegdal, P.L. and Blaskiewicz, R.W. (1984). Evaluation of the potential hazard to Barn Owls of Talon (Brodifacoum bait) used to control rats and house mice. *Environmental Toxicology and Chemistry*, **3**, 167–179.

Henny, C.J. (1969). Geographical variation in mortality rates and production requirements of the barn owl (*Tyto alba*). *Bird-Banding*, **40**, 277–356.

Henttonen, H. (1985). Predation causing extended low densities in microtine cycles: further evidence from shrew dynamics. *Oikos*, **45**, 156–158.

Herrera, C.M. (1974). Regimen alimenticio de *Tyto alba* en Espana suboccidental. *Ardeola*, **19**, 259–394.

Kotler, B.P., Brown, J.S., Smith, R.J. and Wirz II, W.O. (1988). The effects of morphology and body size on rates of owl predation on desert rodents. *Oikos*, **53**, 145–152.

Krebs, J.R., Erichsen, J.T., Webber, M.I. and Charnov, E.L. (1977). Optimal prey selection in the great tit (*Parus major*). *Animal Behaviour*, **25**, 30–38.

Krebs, C.J. and Myers, J.H. (1974). Population cycles in small mammals. *Advances in Ecological Research*, **8**, 267–399.

Krägenow, P. (1970). Die Schleiereule in den Nordbezirken der DDR. *Falke*, **17**, 256–259.

Kulczycki, A. (1964). Study on the make-up of the diet of owls from the Niski Beskid Mts. (Poland). *Acta Zoologica Crakoviensia*, **9**, 529–559.

Lack, D. (1954). *The natural regulation of animal numbers*. Oxford University Press, London.

Lack, D. (1966). *Population studies of birds*. Clarendon Press, Oxford.

Lack, D. (1968). *Ecological adaptations for breeding in birds*. Methuen, London.

Lander, E., Lopez, J., Diaz, C. and Colmenares (1991). Population biology of the barn owl (*Tyto alba*) in Guarico State, Venezuela. *Birds of Prey Bulletin* **4**, 167–173.

Lange, H. (1948). Slorugens (*Tyto alba guttata*, Brehm) Fode, belyst gennem Undersogelser af Gylp. *Dansk Ornitologisk Forenings Tidsskrift*, **42**, 50–84.

Langford, I.K. and Taylor, I.R. (1992). Rates of prey delivery to the nest and chick growth patterns of barn owls *Tyto alba*. In *The Ecology and Conservation of European owls*, eds. C.A. Galbraith, I.R. Taylor and S.M. Percival, pp. 100–103. Joint Nature Conservation Committee, Peterborough, UK.

Laurie, W.A. (1971). The food of the barn owl in the Serengeti National Park, Tanzania, *Journal of the East African Natural History Society*. **28**, 1–4.

Lebreton, J.D., Burnham, K.P., Clobert, J. and Anderson, D.R. (1992). Modelling survival and testing biological hypotheses using marked animals: a unified approach with case studies. *Ecological Monographs*, **62**, 67–118.

Lehmann, U. (1976). Short-term and circadian rhythms in the behaviour of the vole, *Microtus agrestis* (L.). *Oecologia*, **23**, 185–199.

Lehmann, U. and Sommersberg, C.W. (1980). Activity patterns of the common vole, *Microtus arvalis* – automatic recording of behaviour in an enclosure. *Oecologia*, **47**, 61–75.

Lenton, G.M. (1980). Ecology of barn owls *Tyto alba* in the Malay Peninsula with reference to their use in rodent control. Ph.D. Thesis, University of Malaya, Kuala Lumpur.

Lenton, G.M. (1984a). The feeding and breeding ecology of Barn owls *Tyto alba* in Peninsular Malaysia. *Ibis*, **126**, 551–575.

Lenton, G.M. (1984b). Moult of Malayan barn owls *Tyto alba*. *Ibis*, **126**, 188–197.

Lerg, J.M. (1984). Status of the common barn owl in Michigan. *Jack-Pine Warbler*, **62**, 39–48.

Leslie, P.H. (1952). The fertility and population structure of the brown rat (*Rattus novegicus*) in cornricks and some other habitats. *Proceedings of the Zoological Society of London*, **122**, 187–238.

Appendix 3. *A number of sample barn owl diets from Europe and North America to show precise details of species composition*

A. European countries

Prey items eaten (% of total no. eaten)

Species	Canary Is.	Czechoslovakia	Poland	Spain	Italy	Belgium	Holland	Germany	Germany	France Mediterranean	France Burgundy	Ireland	Wales	England	Scotland
Talpa europaea			+				+	+	+	+	+		+	+	+
Sorex araneus		+	14.1			36.8	36.6	18.6	18.6	18.6	18.1	11.9	29.4	23.7	38.3
Sorex minutus			1.4			1.8	1.4	1.3	+		2.0		5.4	4.7	1.4
Neomys fodiens			1.5				+	+	+		+		+	1.3	+
Neomys anomalus				+											
Crocidura leucodon				4.6			+				+				
Crocidura russula				14.9	14.5		10.0	20.2		31.4	12.5				
Crocidura suaveolens					4.8					1.3					
Suncus etruscus				2.4	3.0										
Chiroptera spp.		+	+	+			+		+	+	+			+	
Eliomys quercinus				+					+	+	+				
Glis glis									+	+	+				
Muscardinus avellanarius					2.2			+							
Clethrionomys glareolus			+		10.2	1.2	+	1.6	+		3.0	13.7	4.2	3.3	2.1
Microtus oeconomus			5.6				+								
Microtus agrestis			+			11.9	2.9	16.3	1.7	4.8	4.8		47.8	50.7	44.7
Microtus arvalis		84.2	23.1		33.7	12.4	26.8	25.0	76.1		39.5				
Arvicola terrestris					+		+	2.2	+		+		+	+	
Arvicola sapidus				+						+					
Pitymys subterraneus		+		+		4.3	+				+				
Pitymys				11.9						2.0					
Pitymys duodecimcostatus															
Pitymys savii					26.8										
Micromys minutus			3.5				1.6	+	+				+	3.2	
Apodemus spp.		1.4	4.3	10.8	4.1	4.8	5.0	11.3	1.0	20.7	14.2	55.6	8.3	6.2	12.6
Rattus rattus		1.1	+	+	+		+			1.8					
Rattus norvegicus			+				+					9.0	1.0		
Rattus spp.	4.1								+					3.2	+
Mus spp.	70.3	1.5	26.4	45.0	2.8	2.8	2.3	+	+	17.3	+	5.8	+	+	
Mustela nivalis							+	+	+		+			+	
Aves	2.2		13.2	4.3	2.0	2.9	2.0	+	+	10.5	+	2.2	1.1	2.3	+
Reptilia	2.5			+											
Amphibia	+		1.9	4.3	2.0		+	+	+	6.8	1.1	+	+		+
Insecta	18.3		+	4.3	2.0		+			1.0	+	+	+		
Total sample size (No. of items)	2058	913	16944	14801	3661	34866	92383	33469	11913	18561	40614	8229	4575	16350	40294
Reference	Martin *et al.* 1985	Pikula *et al.* 1984	Ruprecht 1964	Herrera 1974	Lovari *et al.* 1976	Straeten and Asselberg 1973	De Bruin 1979	Bethge and Hayo 1979	Bohnsack 1966	Libois *et al.* 1983	Baudvin 1983	Smal 1987	Glue 1974	Buckley and Goldsmith 1975	Taylor this study

Appendix 3 (*cont.*)
B. North American areas

Species	Prey items eaten (% of total no. eaten)								
	Texas	Oklahoma	Virginia	New Jersey	Ohio	Colorado	Utah	Idaho	British Columbia, Canada
Shrews, *Sorex* spp.				+	+		4.7	+	11.7
Masked shrew *S. cinereus*									1.5
Vagrant shrew *S. vagrans*									3.1
Short-tailed shrew *Blarina brevicauda*	16.3	+	13.0	7.6	20.2	+			
Least shrew *Cryptotis parva*			6.0	+	+				1.0
Moles Talpidae					+				
Bats Chiroptera					+		+	+	+
Cotton tails *Sylvilagus* spp.	2.7		+		+	1.5			
Eastern chipmunk *Tamias striatus*									
Pocket gophers *Thomomys* spp.	4.0	1.1				7.3	+	5.6	
Pocket mice *Perognathus* spp.	9.6	42.1				7.3	+	6.6	
Kangaroo rat *Dipodomys* spp.						4.0		8.2	
Rice rat *Oryzomys* spp.	5.8	18.5	1.0	4.2		7.3	6.4	3.5	
Harvest mice *Reithrodontomys* spp.	10.2	14.1	2.0			25.1	7.7	12.8	
White-footed mice *Peromyscus* spp.	1.8	8.4	1.0	2.0	5.0	+		+	1.7
Grasshopper mice *Onychomys* spp.		11.8				+			
Cotton rat *Sigmodon* spp.	10.8	3.9	+						
Wood rat *Neotoma* spp.	4.7					+		+	
Meadow vole *Microtus pennsylvanicus*			55.0	74.1	63.9	9.5			
Mountain vole *M. montanus*							72.0	55.3	
Townsend's vole *M. townsendii*									73.0
Oregon vole *M. oregoni*									1.3
Prairie vole *M. ochrogaster*					+	34.8			
Voles *Microtus* spp.				+	+	+	+	+	
Muskrat *Ondatra*									
Ship rat *Rattus rattus*									1.0
Norway rat *R. norvegicus*			2.0	1.9	+		+		1.3
House mouse *Mus musculus*			12.0	6.7	+	+	5.3	6.9	1.1
Jumping mouse *Zapus hudsonius*				1.1	4.1				
Pygmy mouse *Baiomys taylori*	19.7								
Aves	12.9		6.0	1.3	1.5	1.4	2.5	1.5	1.9
Reptilia									
Amphibia									
Insecta	1.7						+		
Number of items	11408	680	1061	9531	12589	4366	25330	38199	30218
Reference	Otteni *et al.* 1972	Ault 1971	Rosenburg 1986	Colvin 1984	Colvin and McLean 1986	Marti 1974	Marti 1988	Marti 1988	Campbell *et al.* 1987

+ denotes values for the prey eaten of less than 1% of the total number.

Location	Latitude												Reference
Spain	42°N	96.8	79.2	10	*Mus* spp.	44.4	*Mus* spp.	37.3	4	4	4–30	17	Brunet-Lecompte and Delibes 1984
Italy	38°N	87.1	68.8	12	*Mus musculus*	44.9	*Mus musculus*	37.0	4	4	10–30	20	Herrera 1974
Italy	43°N	96.1	77.4	12	*Apodemus* spp.	33.7	*Apodemus* spp.	49.0	5	3	5–80	18	Lovari *et al.* 1976
Palestine	32°N	77.7	65.0	11	*M. guntheri*	46.1	*M. guntheri*	–	3	–	–	–	Dor 1947
Canary Is	28°N	74.5	74.5	2	*Mus musculus*	70.3	*Mus musculus*	76.9	2	2	1–60	17	Martin *et al.* 1985
Morocco	32°N	99.7	88.7	4	*Mus musculus*	69.7	*Mus musculus*	–	4	–	–	–	Saint-Girons and Thouy 1978
North and South America													
British Columbia	48°N	97.0	80.1	18	*M. townsendii*	73.0	*M. townsendii*	82.5	2	1	6–65	55	Campbell *et al.* 1987
Washington	46°N	97.8	97.8	9	*Perognathus parvus*	63.9	*Thomomys talpoides*	39.3	3	3	10–100	28	Knight and Jackman 1984
Oregon	45°N	99.7	57.2	10	*S. vagrans*	42.3	*S. townsendii*	71.6	3	2	6–60	29	Giger 1965
Idaho	43°N	98.5	93.5	15	*M. montanus*	55.3	*M. montanus*	55.7	4	3	10–70	34	Marti 1988
Utah	41°N	97.0	92.6	18	*Microtus* spp. (*montanus* and *pennsylvanicus*)	72.0	*Microtus* spp.	84.4	3	2	10–50	38	Marti 1988
Michigan	43°N	98.9	92.0	13	*M. pennsylvanicus*	85.0	*M. pennsylvanicus*	93.0	1	1	15–40	38	Wallace 1948
Ohio	41°N	98.5	74.5	17	*M. pennsylvanicus*	63.9	*M. pennsylvanicus*	75.8	2	2	18–90	33	Colvin and McLean 1986
New Jersey	39°N	98.7	89.0	11	*M. pennsylvanicus*	74.1	*M. pennsylvanicus*	80.4	2	1	15–150	35	Colvin 1984
Virginia	38°N	94.0	76.0	11	*M. pennsylvanicus*	55.0	*M. pennsylvanicus*	60.0	3	3	5–65	42	Rosenburg 1986
Colorado	40°N	98.6	97.0	13	*M. ochrogaster*	34.8	*M. ochrogaster*	30.1	5	6	8–130	46	Marti 1974
Oklahoma	36°N	100	100	9	*Perognathus* spp.	42.0	*Perognathus* spp.	29.2	4	4	18–120	55	Ault 1971
California	37°N	98.5	98.5	7	*Perognathus inornatus*	44.6	*Thomomys bottae*	71.4	2	2	20–100	–	Fitch 1947
Georgia, Alabama, S. Carolina	33°N	96.5	78.9	14	*Sigmodon hispidus*	54.8	*Sigmodon hispidus*	–	4	1	20–115	76 (?)	French and Wharton 1975
Texas	29°N	92.0	82.5	13	*Sigmodon hispidus*	34.2	*Sigmodon hispidus*	69.8	7	3	2–115	58	Byrd 1982
Texas	28°N	85.4	66.6	10	*Baiomys taylori*	19.7	*Sigmodon hispidus*	28.0	8	6	5–120	65	Otteni *et al.* 1972

Appendix 2 (*cont.*)

Country	Latitude	Diet content (% by number)		Total number terrestrial small mammal species	Most frequent prey species (by number)		Most important prey species (by biomass)		No. of species forming 80% of diet by		Prey weight (g)		Source
		Small mammals	Rodents		Species	% of diet	Species	% of diet	Number	Biomass	Normal range	Average	
Chile		95.1	90.1	9	Oryzomys longicaudatus	28	Oryzomys longicaudatus	?	5	?	17–70	60	Herrera and Jaksic 1980
Africa													
Transvaal		92.9	85.8	25	Praomys natalensis	45.9	Praomys natalensis	45.8	6	2	7–110	42	Vernon 1972
Cape Province		91.6	85.6	20	Gerbillus paeba	28.3	Desmodillus auricularis	25.1	6	10	10–110	39	Vernon 1972
	30°S	96.8	90.0	13	Praomys natalensis	21.0	Otomys irroratus	29.0	6	5	7–110	63	Perrins 1982
Natal		96.9	81.2	12	Praomys natalensis	58.0	Praomys natalensis	60.4	5	2	7–110	42	Vernon 1972
Namibia	25°S	89.5	85.2	11	Gerbillus and Tatera spp.	74.9	Gerbillus and Tatera spp.	63.0	3	6	7–50	27	Vernon 1972
Australia													
Victoria		99.1	97.8	4	Mus musculus	96.3	Mus musculus	80	1	1	15–80	20	Morton 1975
NSW and S. Australia	28°S	98.0	94.9	10	Mus musculus	82.8	Mus musculus	64.5	1	2	15–100	19	Morton and Martin 1979
S.W. Queensland	25°S	99.3	94.8	10	Rattus villosissimus	45.1	Rattus villosissimus	75.6	4	2	20–100	60	Morton et al. 1977
Northern Territory	20°S	85.5	80.3	10	Notomys alexis	41.1	Notomys alexis	54.5	7	5	10–30	23	Smith and Cole 1989
Other areas													
Malaysia	2°N	99.2	98.2	8	Rattus tiomanicus	89.8	Rattus tiomanicus	90	1	1	–	80	Lenton 1984a
Galapagos Islands	1°S	44.9	44.9	2	Mus musculus	39.3	Mus musculus	47.6	2	2	2–60	25	de Groot 1982

Libois, R.M., Fons, R. and Saint-Girons, M.C. (1983). Le régime alimentaire de la Chouette effraie, *Tyto alba*, dans les Pyrénées-orientales. Etude des variations écogéographiques. *Revue Ecology* (Terre Vie), **37**, 187–217.

Lidicker, W.Z. (1988). Solving the enigma of microtine 'cycles'. *Journal of Mammalogy*, **69**, 225–235.

Lockie, J.D. (1955). The breeding habits and food of the short-eared owl after a vole plague. *Bird Study*, **2**, 53–69.

Lofts, B. and Murton, R.K. (1968). Photoperiodic and physiological adaptations regulating avian breeding cycles and their ecological significance. *Journal of Zoology* (London), **155**, 327–394.

Looman, S.J. (1985). *Productivity, food habits and sexing of barn owls in Utah*. M.Sc. Thesis, Brigham Young University, Provo, Utah.

Lovari, S., Renzoni, A. and Fondi, R. (1976). The predatory habits of the barn owl (*Tyto alba*) in relation to vegetation cover. *Bollettino di Zoologia*, **43**, 173–191.

Lynch, C.D. (1983). The mammals of the Orange Free State. *Memoirs van die Nasionale Museum Bloemfontein*, **18**, 20–32.

MacArthur, R.H. and Pianka, E.R. (1966). On optimal use of a patchy environment. *American Naturalist*, **100**, 603–609.

MacArthur, R.H. and Wilson, E.O. (1967). *The theory of island biogeography*. Princeton University Press, Princeton, NJ.

Maerks, H. (1954). Uber den Einfluss der Witterung auf den Massenwechsel der Feldmaus (*Microtus arvalis* Pallas) in der Wesermarsch. *Nachrichtenbl Deutsch Pflanzenschutzdienst*, **6**, 101–108.

Marshall, J.D., Hager, C.H. and McKee, G. (1986). The barn owl egg: weight loss characteristics, fresh weight prediction and incubation period. *Raptor Research*, **20**, 108–112.

Marten, G.G. (1973). Time patterns of *Peromyscus* activity and their correlations with weather. *Journal of Mammalogy*, **54**, 169–188.

Marti, C.D. (1973). Ten years of barn owl prey data from a Colorado nest site. *Wilson Bulletin*, **85**, 85–86.

Marti, C.D. (1974). Feeding ecology of four sympatric owls. *Condor*, **76**, 45–61.

Marti, C.D. (1988). A long-term study of food-niche dynamics in the common barn owl: comparisons within and between populations. *Canadian Journal of Zoology*, **66**, 1803–1812.

Marti, C.D. (1990). Sex and age determination in the barn owl and a test of mate choice. *Auk*, **107**, 246–254.

Marti, C.D. and Wagner, P.W. (1985). Winter mortality in common barn owls and its effect on population density and reproduction. *Condor*, **87**, 111–115.

Martin, A., Emmerson, K. and Ascanio, M. (1985). Regimen alimenticio de *Tyto alba* (Scopoli 1769) en la Isla de Tenerife (Islas Canarias). *Ardeola*, **32**, 9–15.

Martin, G.R. (1977). Absolute visual threshold and scotopic spectral sensitivity in the Tawny Owl, *Strix aluco. Nature* (London), **268**, 636–638.

Martin, G.R. (1982). An owl's eye: schematic optics and visual performance in *Strix aluco* L. *Journal of Comparative Physiology*, **145**, 341–349.

Maser, C. and Brodie, E.D. (1966). A study of owl pellet contents from Linn, Benton and Polk Counties, Oregon. *Murrelet*, **47**, 9–14.

Masman, D. (1986). The annual cycle of the kestrel, *Falco tinnunculus*, a study in behavioural energetics. Ph.D. Thesis, University of Groningen.

Masman, D., Daan, S. and Beldhuis, J.A. (1988). Ecological energetics of the kestrel: daily energy expenditure throughout the year based on the time-budget, food intake and doubly labelled water methods. *Ardea*, **76**, 64–81.

Masman, D. and Daan, S. (1987). The allocation of energy in the annual cycle of the kestrel, *Falco tinnunculus*. In *The ancient kestrel*, eds. D.M. Bird and R. Bowman, pp. 124–136. Raptor Research Foundation, Quebec.

Masman, D., Gordijn, M., Daan, S. and Dijkstra, C. (1986). Bioenergetics of the kestrel. *Falco tinnunculus*: the variability of natural intake rates. *Ardea*, **74**, 24–38.

Meijer, T. (1980). Reproductive decisions in the kestrel *Falco tinnunculus*. Ph.D. Thesis, University of Groningen.

Mendenhall, V.M. and Pank, L.F. (1980). Secondary poisoning of owls by anticoagulant rodenticides. *Wildlife Society Bulletin*, **8**, 311–315.

Mendenhall, V.M., Klaas, E.E. and McLane, M.A.R. (1983). Breeding success of barn owls (*Tyto alba*) fed low levels of DDE and Dieldrin. *Archives in Environmental Contamination and Toxicology*, **12**, 235–240.

Middleton, A.D. (1930). Cycles in the numbers of British voles (*Microtus*). *Journal of Ecology*, **18**, 156–165.

Middleton, A.D. (1931). A further contribution to the study of British voles (*Microtus*). *Journal of Ecology*, **19**, 190–199.

Mikkola, H. (1983). *Owls of Europe*. T. and A.D. Poyser, Carlton, UK.

Morris, P. (1979). Rats in the diet of the barn owl. *Journal of Zoology*, **189**, 540–545.

Morton, S.R. (1975). The diet of the barn owl, *Tyto alba*, in southern Victoria. *Emu*, **75**, 31–34.

Morton, S.R., Happold, M., Lee, A.K., Lee, I. and McMillen, R.E. (1977). The diet of the barn owl *Tyto alba* in south-western Queensland. *Australian Wildlife Research*, **4**, 91–97.

Morton, S.R. and Martin, A.A. (1979). Feeding ecology of the barn owl *Tyto alba*, in arid southern Australia. *Australian Wildlife Research*, **6**, 191–204.

Mourer-Chauvire, C., Alcover, J.A., Moya, S. and Pons, J. (1980). Une nouvelle forme insulaire d'effraie géante, *Tyto balearica* n. sp. du Plio-Pleistocene des Baléares. *Géobios*, **13**, 803–811.

Mueller, H.C. and Berger, D.D. (1959). Some long distance barn owl recoveries. *Bird Banding*, **30**, 182.

Muller, Y. (1989). Fluctuations d'abondance de la Chouette effraie *Tyto alba* en Alsace-Lorraine de 1977 à 1988. *Aves*, **26**, 131–141.

Muller, Y. (1990). Chevauchement des nidifications successives chez la Chouette effraie (*Tyto alba*). *Alauda*, **58**, 217–220.

Muller, Y. (1991). *Les secondes nichées chez la Chouette effraie* (Tyto alba). *Rapaces Nocturnes*. Actes due 30 Colloque interrégional d'ornithologie Porrentruy (Suisse), pp. 173–188.

Mumford, R.E. and Keller, C.E. (1984). *The Birds of Indiana*. Indiana University Press. Bloomington, IN.

Mutze, G.J., Veitch, L.G. and Miller, R.B. (1990). Mouse plagues in south Australian cereal growing areas. II. An empirical model for prediction of plagues. *Australian Wildlife Research*, **17**, 313–324.

Myllimaki, A., Christiansen, E. and Hansson, L. (1977). Five year surveillance of small mammal abundance in Scandinavia. *European and Mediterranean Plant Protection Organization Bulletin*, **7**, 385–396.

Nature Conservancy Council (1984). *Nature Conservation in Great Britain*. NCC, Shrewsbury, UK.

Neal, B.R. (1968). The ecology of small rodents in the grassland community of the Queen Elizabeth National Park, Uganda. Ph.D. Thesis, University of Southampton.

Neal, B.R. (1970). The habitat distribution and activity of a rodent population in western Uganda, with particular reference to the effects of burning. *Revue de Zoologie et de Botanique Africaines*, **81**, 29–50.

Newsome, A.E. (1969a). A population of house mice temporarily inhabiting a South Australian wheatfield. *Journal of Animal Ecology*, **38**, 341–359.

Newsome, A.E. (1969b). A population study of house mice permanently inhabiting a reed-bed in South Australia. *Journal of Animal Ecology*, **38**, 361–377.

Newsome, A.E. and Corbett, L.K. (1975). Outbreaks of rodents in semi-arid and arid Australia: causes, preventions and evolutionary considerations. In *Rodents in Desert Environments* eds. I. Prakash and P.K. Gosh, pp. 117–153. Junk, The Hague.

Newsome, A.E. and Crowcroft, W.P. (1971). Outbreaks of house mice in South Australia. *CSIRO Wildlife Research*, **16**, 111–120.

Newton, I. (1967). Feather growth and moult in some captive finches. *Bird Study*, **14**, 10–24.

Newton, I. (1979). *Population Ecology of Raptors*. T. and A.D. Poyser, Berkhamsted, UK.

Newton, I. (1986). *The Sparrowhawk*. T. and A.D. Poyser, Carlton.

Newton, I. and Marquiss, M. (1981a). Moult in the Sparrowhawk. *Ardea*, **70**, 163–172.

Newton, I. and Marquiss, M. (1981b). Effect of additional food on laying dates and clutch sizes of Sparrowhawks. *Ornis Scandinavica*. **12**, 224–229.

Newton, I. and Marquiss, M. (1984). Seasonal trend in the breeding performance of Sparrowhawks. *Journal of Animal Ecology*, **53**, 809–829.

Newton, I., Wyllie, I. and Asher, A. (1991). Mortality causes in British barn owls *Tyto alba* with a discussion of aldrin-dieldrin poisoning. *Ibis*, **133**, 162–169.

Newton, I., Wyllie, I. and Freestone, P. (1990). Rodenticides in British barn owls. *Environmental Pollution*, **68**, 101–117.

Nishimura, K. and Abe, M.T. (1988). Prey susceptibilities, prey utilisation and variable attack efficiencies of Ural owls. *Oecologia*, **77**, 414–422.

Otteni, L.C., Bolen, E.G. and Cottam, C. (1972). Predator-prey relationships and reproduction of the Barn Owl in southern Texas. *Wilson Bulletin*, **48**, 434–448.

Parkes, K.C. and Phillips, A.R. (1978). Two new Caribbean subspecies of Barn Owl (*Tyto alba*), with remarks on variation in other populations. *Annals of Carnegie Museum*, **47**, 479–492.

Payne, R.S. (1971). Acoustic location of prey by barn owls *Tyto alba*. *Journal of Experimental Biology*, **54**, 535–573.

Payne, R.S. and Drury, W.H. (1958). Marksman of the darkness. *Natural History* (NY), **67**, 316–323.

Peakall, D.B. (1970). pp-DDT: effect on calcium metabolism of extradiol in the blood. *Science* (NY), **168**, 592–594.

Pearson, T.H. (1968). The feeding biology of seabird species breeding on the Farne Is., Northumberland. *Journal of Animal Ecology*, **37**, 521–552.

Percival, S.M. (1990). Population trends in British barn owls, *Tyto alba* and Tawny owls *Trix aluco*, in relation to environmental change. *British Trust for Ornithology Research*, Report No. 57, pp. 129.

Perrins, C.M. (1970). The timing of birds' breeding seasons. *Ibis*, **112**, 242–255.

Perrins, C.M. and Moss, D. (1975). Reproductive rates in the great tit. *Journal of Animal Ecology*, **44**, 695–706.

Perrins, M.R. (1982). Prey specificity of the barn owl, *Tyto alba*, in the Great Fish River valley of the Eastern Cape Province. *South African Journal of Wildlife Research* **12**, 14–25.

Peters, R.H. (1983). *The ecological implications of body size*. Cambridge University Press, Cambridge.

Pettifor, R.A. (1983). Seasonal variation and associated energy implications in the hunting behaviour of the kestrel. *Bird Study* **30**, 201–206.

Pidoplitschka, J.G. (1937). Ergebnisse der Gewölluntersuchungen in den Jahren 1924–1935. *Verobb. Instituts Fuer Zoologischen und Biologischen* (Ukraine), **19**, 101–170.

Piechocki, R. von (1960). Über die Winterverlustre der Schleiereule (*Tyto alba*). *Vogelwarte*, **20**, 274–280.

Piechocki, R. von (1974). Über die Grossgefieder-Mauser eines gekafigten Paares der Schleiereule (*Tyto alba*). *Journal of Ornithology* **115**, 436–444.

Pietiainen, H., Saurola, P. and Kolunen, H. (1984). The reproductive constraints on moult in the Ural Owl *Strix uralensis*. *Annals Zoologica Fennici*, **21**, 277–281.

Pijanowski, B.C. (1992). A revision of Lack's brood reduction hypothesis. *American Naturalist* **139**, 1270–1292.

Pikula, J., Beklova, M., Kubik, V. (1984). The breeding bionomy of *Tyto alba*. *Acta Scientiarum Naturalium, Academiae Scientiarum Bohemoslovacae Brno*, **18**, 1–53.

Prestt, I. (1965). An enquiry into the recent breeding status of some smaller birds of prey and crows in Britain. *Bird Study* **12**, 196–221.

Pricam, R. and Zelenka, G. (1964). Le régime alimentaire de la Chouette effraie *Tyto alba* sur la rive gauche du Léman. *Alanda*, **32**, 175–195.

Pyke, G.M., Pulliam, H.R. and Charnov, E.L. (1977). Optimal foraging: a selective review of theory and tests. *Quarterly Review of Biology*, **52**, 137–154.

Quine, D.B. and Konishi, M. (1974). Absolute frequency discrimination in the barn owl. *Journal of Comparative Physiology*, **93**, 347–360.

Raczynski, J. and Ruprecht, A.L. (1974). The effect of digestion on the osteological composition of the owl pellets. *Acta Ornithologica*, **15**, 1–12.

Rahn, H. and Ar, A. (1974). The avian egg: incubation time and water loss. *Condor*, **76**, 147–152.

Rahn, H., Paganelli, C.V. and Ar, A. (1975). Relation of avian egg to body weight. *Auk*, **92**, 698–705.

Raptor Group RUG/RIJP (1982). Timing of vole hunting in aerial predators. *Mammal Review*, **12**, 169–181.

Reese, J.G. (1972). A Chesapeake barn owl population. *Auk*, **89**, 106–114.

Regisser, B. (1991). *Bilan de 13 années de protection et de suivi de la Chouette effraie,* Tyto alba, *dans le Haut-Rhin (France) de 1978 à 1990.* Rapace Nocturnes, Actes du 30 Colloque interrégional d'ornithologie Porrentruy (Suisse).

Rice, W.R. (1982). Acoustic location of prey by the Marsh Hawk: adaptation to concealed prey. *Auk*, **99** 403–413.

Ricklefs, R.E. (1968). Patterns of growth in birds. *Ibis*, **110**, 419–451.

Ritter, F. and Gorner, M. (1977). Untersuchungen uber die Beziehung zwischen Futterungsaktivitat und Beutetierzahl bei der Schleireule. *Der Falke*, **24**, 344–348.

Rosenburg, C.P. (1986). Barn Owl Habitat and Prey Use in Agricultural Eastern Virginia. M.Sc. Thesis, College of William and Mary, Williamsburg, VA.

Rowe, F.P., Taylor, E.J. and Chudley, A.H.J. (1963). The numbers and movements of house-mice (*Mus musculus*, L.) in the vicinity of four corn ricks. *Journal of Animal Ecology*, **32**, 87–97.

Rowe, F.P., Taylor, E.J. and Chudley, A.H.J. (1964). The effect of crowding on the reproduction of the house-mouse (*Mus musculus*, L.) living in corn-ricks. *Journal of Animal Ecology*, **33**, 477–483.

Ruprecht, A.L. (1979a). Bats (Chiroptera) as constituents of the food of barn owls (*Tyto alba*) in Poland. *Acta Ornithologica*, **14**, 25–38.

Ruprecht, A.L. (1979b). Food of the Barn Owl, *Tyto alba guttata* from Kujawy. *Acta Ornithologica*, **26**, 493–511.

Ruprecht, A.L. (1964). Analyse der Nahrungsbestandteile der Schlkeiereule, *Tyto alba guttata* vorkommend in Aleksandrow Koj. Ciechocinek und Raciazek in den Jahren 1960–1961. *Zeszyty Naukowe Uniwersytetu Mikolaja Kopernika w Toruniu, Biologia*, **7**, 45–66.

Saint-Girons, M.C. (1965). Notes sur les mammifères de France IV. Prévélements exercises sur des populations de petits mammiferes par la Chouette effraie *Tyto alba* (Région de Lyon). *Mammalia*, **29**, 42–53.

Saint-Girons, M.C. and Thouy, P. (1978). Fluctéations dans les populations de Souris (*Mus spretus*) en région méditerranéenne. *Bulletin d'Ecologie* **9**, 211–218.

Sandell, M., Astrom, M., Atlegrim, O., Danell, K., Edenius, L., Hjalten, J., Lundberg, P., Palo, T., Pettersson, R. and Sjoberg, G. (1991). 'Cyclic' and 'non-cyclic' small mammal populations: an artificial dichotomy. *Oikos*, **61**, 281–284.

Saurola, P. (1989). Breeding strategy of the Ural Owl *Strix uralensis*. In *Raptors in the Modern World*, eds. B.-U. Meyburg and R.D. Chancellor, pp. 235–241. WWGPB, Berlin.

Sauter, U. (1956). Beiträge zur ökologie der Schleiereule (*Tyto alba*) nach den ringfunden. *Die Vogelwarte*, **18**, 109–151.

Schifferli, A. (1957). Aler und Sterblichkeit bei Waldkauz (*Strix aluco*) und Schleiereule (*Tyto alba*) in der Schweiz. *Ornithologische Beobachter*, **54**, 50–56.

Schmidt, E. (1973). Die Nahrung der Schleiereule (*Tyto alba*) in Europa. *Angewandle Zoologie*, **60**, 43–70.

Schmidt, E. and Szlivka, L. (1968). Adatok a reti fulesbagoly (*Asio flammeus*) teli tapalkozasahoz a Bacskaban (Eszak-Jugoszlavia). *Aquila*, **75**, 227–229.

Schnider, U. (1972). Massenwechsel der ermause, *Microtus agrestis* in Sud-Neidersachen von 1952–1971. *Zeitschrift für Angewandte Zoologie*, **59**, 189–205.

Schodde, R. and Mason, I. (1980). *Nocturnal Birds of Australia*. Lansdowne, Melbourne.

Schoener, T.W. (1968). Size of feeding territories among birds. *Ecology*, **49**, 123–141.

Schonfeld, M. (1974). Ringfundaus-wertung der 1964–1972 in er DDR beringet Schleiereulen *Tyto alba guttata* Brehm. *Jahrbuch Vogelwarte Hiddensee*, **4**, 90–122.

Schonfeld, M. and Girbig, G. (1975). Beiträge zur Brutbiologie der Schleiereule, *Tyto alba*, besonders unter Berucksichtgung der Feldmausdichte. *Hercynia*, **12**, 257–319.

Schonfeld, M., Girbig, G. and Strum, H. (1977). Beiträge zur Populationsdynamik der Schleiereule *Tyto alba*. *Hercynia*, **14**, 303–351.

Schonfeld, M. and Piechocki, R. (1974). Beiträge zur Grossgefiedermauser der Schleiereule, *Tyto alba guttata*. *Journal of Ornithology*, **115**, 418–435.

Seel, D.C., Thomson, A.G. and Turner, J.C.E. (1983). *Distribution and breeding of the barn owl* Tyto alba *on Anglesey, North Wales*. Bangor Occasional Paper No. 16. Institute of Terrestrial Ecology, Bangor.

Sharrock, J.T.R. (1976). *The atlas of Breeding Birds in Britain and Ireland* T. & A.D. Poyser, Berkhamsted, UK.

Shawyer, C. (1987). *The Barn Owl in the British Isles – its Past, Present and Future*. The Hawk Trust, London.

Shutt, L. and Bird, D.M. (1985). Influence of nestling experience on nest-type selection in captive kestrels. *Animal Behaviour*, **33**, 1028–1031.

Slagsvold, T. and Lifjeld, J.T. (1989). Hatching asynchrony in birds: the hypothesis of sexual conflict over parental investment. *American Naturalist*, **134**, 239–253.

Smal, C.M. (1987). The diet of the Barn Owl *Tyto alba* in southern Ireland, with reference to a recently introduced prey species – the Bank Vole *Clethrionomys glareolus*. *Bird Study*, **34**, 113–125.

Smal, C.M., Halim, A.H. and Amiruddin, M.D. (1990). Predictive modelling of rat populations in relation to use of rodenticides or predators for control. *Palm Oil Research Institute of Malaysia*, occasional paper, Kuala Lumpur.

Smith, C.R. and Richmond, M.E. (1972). Factors affecting pellet egestion and gastric pH in the barn owl. *Wilson Bulletin*, **84**, 179–186.

Smith, D.G., Wilson, C.R. and Frost, H.H. (1972). Seasonal food habits of barn owls in Utah. *Great Basin Naturalist*, **31**, 229–234.

Smith, D.G. and Marti, C.D. (1976). Distributional status and ecology of barn owls in Utah. *Raptor Research*, **10**, 33–44.

Smith, J.D.B. and Cole, J. (1989). Diet of the barn owl *Tyto alba*, in the Tanami Desert, Northern Territory. *Australian Wildlife Research*, **16**, 611–624.

Soucy, L. J. Jr. (1985). Bermuda recovery of a barn owl banded in New Jersey. *Journal of Field Ornithology*, **56**, 274.

Southern, H.N. (1954). Tawny owls and their prey. *Ibis*, **96**, 384–410.

Southern, H.N. (1970). The natural control of a population of tawny owls, *Strix aluco*. *Journal of Zoology* (London), **162**, 197–285.

Southern, H.N. and Lowe, V.P.W. (1982). Predation by tawny owls (*strix aluco*) on bank voles (*Clethrionomys glareolus*) and wood mice (*Apodemus sylvaticus*). *Journal of Zoology* (London), **198**, 83–102.

Southwick, C.H. (1958). Population characteristics of house mice living in English corn ricks: density relationships. *Proceedings of the Zoological Society of London*, **131**, 163–175.

Spencer, A.W. and Pettus, D. (1966). Habitat preferences in five sympatric species of long-tailed shrews. *Ecology*, **47**, 677–683.

Stearns, S.C. (1992). *The Evolution of Life Histories*. Oxford University Press, Oxford.

Stewart, P.A. (1952). Dispersal, breeding behaviour, and longevity of banded barn owls in North America. *Auk*, **69**, 227–245.

Stewart, P.A. (1980). Population trends of barn owls in North America. *American Birds* **34**, 698–700.

Straeten, E. van der and Asselberg, R. (1973). Het voedsel van de kerkuil in Belgie. *Giervalk*, **63**, 149–159.

Straka, F. and Gerasimov, S. (1971). Correlations between some climatic factors and the abundance of *Microtus arvalis* in Bulgaria. *Annales Zoological Fennici*, **8**, 113–116.

Stresemann, E. and Stresemann, V. (1966). Die Mauser der Vogel. *Journal of Ornithology*, **107** (Sonderheft) Striges, 357–375.

Summerhayes, V.S. (1941). The effect of voles on vegetation. *Journal of Ecology*, **29**, 14–48.

Taitt, M.J. and Krebs, C.J. (1985a). Predation, cover and food manipulations during a spring decline of *Microtus townsendii*. *Journal of Animal Ecology*, **52**, 837–848.

Taitt, M.J. and Krebs, C.J. (1985b). Population dynamics and cycles. In *Biology of New World Microtus*, Ed. R.H. Tamarin, pp. 567–620. *American Society of Mammalogy*, Special Publication vol. 8.

Tapper, S. (1979). The effect of fluctuating vole numbers (*Microtus agrestis*) on a population of weasels (*Mustela nivalis*) on farmland. *Journal of Animal Ecology*, **48**, 603–617.

Tate, J. (1986). The blue list for 1986. *American Birds* **40**, 227–236.

Taylor, I.R. (1991a). Effects of nest inspections and radiotagging on barn owl breeding success. *Journal of Wildlife Management*, **55**, 312–315.

Taylor, I.R. (1991b). *Sélection du site de nid et taux de pertes de ces sites chez la Chouette effraie*, Tyto alba, *en Ecosse*. Rapaces Nocturnes, Actes 30 Colloque interrégional d'ornithologie, Porrentruy (Suisse), pp. 255–266.

Taylor, I.R. (1992). An assessment of the significance of annual fluctuations in snow cover in determining short-term population changes in field vole *Microtus agrestis* and barn owl *Tyto alba* populations in Britain. In *The Ecology and Conservation of European Owls*, eds. C.A. Galbraith, I.R. Taylor and C.M. Percival, pp. 32–38. Joint Nature Conservation Committee, Peterborough, UK.

Taylor, I.R., Dowell, A., Irving, T., Langford, I.K. and Shaw, G. (1988). The distribution and abundance of the barn owl *Tyto alba* in southwest Scotland. *Scottish Birds*, **15**, 40–43.

Taylor, I.R., Dowell, A. and Shaw, G. (1992). The population ecology and conservation of barn owls *Tyto alba* in coniferous plantations. In *The Ecology and Conservation of European Owls*, eds. C.A. Galbraith, I.R. Taylor and C.M. Percival, pp. 16–21. Joint Nature Conservation Committee, Peterborough, UK.

Taylor, I.R. and Massheder, J. (1992). The dynamics of depleted and introduced barn owl, *Tyto alba*, population: a modelling approach. In *The Ecology and Conservation of European Owls*, eds. C.A. Galbraith, I.R. Taylor and C.M. Percival, pp. 104–109. Joint Nature Conservation Committee, Peterborough, UK.

Taylor, K.D. (1968). An outbreak of rats in agricultural areas of Kenya. *East African Agriculture and Forestry Journal*, **34**, 66–77.

Thorpe, W.H. and Griffin, D.R. (1962). The lack of ultrasonic components in the flight noise of owls compared with other birds. *Ibis*, **104**, 256–257.

Ticehurst, C.B. (1935). On the food of the Barn Owl and its bearing on the Barn Owl population. *Ibis*, **2**, 329–335.

Tomich, P. (1971). Notes on the feeding behaviour of raptorial birds in Hawaii. *Elepaio*, **31**, 111–114.

Trivers, R.L. (1972). Parental investment and sexual selection. In *Sexual Selection and the Descent of Man*, ed. B. Campbell, pp. 136–137. Aldine, Chicago.

Turner, B.N. and Iverson, S.L. (1973). The annual cycle of aggression in male *Microtus pennsylvanicus* and its relation to population parameters. *Ecology*, **54**, 967–981.

Valente, A. (1981). Vertebrate prey in pellets of the barn owl, *Tyto alba*, from Planet Downs Station, south-western Queensland. *Australian Wildlife Research*, **8**, 181–185.

Vargas, J.M., Palomo, L.J. and Palmquist, P. (1988). Predacion y seleccion interspecifica de la lechuza comun (*Tyto alba*) sobre el raton moruno (*Mus spretus*). *Ardeola*, **35**, 109–123.

Verheyen, R.F. (1969). La mortalité de Chouette effraie en Bélgique. *Bulletin du Réserve Naturelles Ornithologie de Belgique, 1968/69*, **49**, 49–80.

Vernon, C.J. (1972). An analysis of owl pellets collected in southern Africa. *Ostrich*, **43**, 109–124.

Vezina, A.F. (1985). Empirical relationships between predator and prey size among terrestrial vertebrate predators. *Oecologia*, **67**, 555–565.

Vickery, W.L. and Bider, J.R. (1981). The influence of weather on rodent activity. *Journal of Mammalogy*, **62**, 140–145.

Village, A. (1983). Seasonal changes in the hunting behaviour of kestrels. *Ardea*, **71**, 117–124.

Village, A. (1990). *The Kestrel*. T and A.D. Poyser, Carlton, UK.

Voitkevitch, A.A. (1966). *The Feathers and Plumage of Birds* Sidgwick and Jackson, London.

Wallace, G.J. (1948). *The Barn Owl in Michigan*. Michigan State College, Agricultural Experiment Station, Technical Bulletin No. 208.

Wallick, L.G. and Barrett, G.W. (1976). Bioenergetics and prey selection of captive barn owls. *Condor*, **78**, 139–141.

Webster, J.A. (1973). Seasonal variation in mammal contents of barn owl castings. *Bird Study*, **20**, 185–196.

Wells, D.R. (1972). Bird Report: 1969. *Malayan Nature Journal*, **25**, 43–61.

West, G.C. (1968). Bioenergetics of captive Willow Ptarmigan under natural conditions. *Ecology*, **49**, 1035–1045.

Wetmore, A. (1968). The birds of the Republic of Panama. Part 2. *Smithsonian Miscellaneous Collections*, **150**, 1–605.

Wijnandts, H. (1984). *Ecological energetics of the long-eared owl (Asio otus)*. Ph.D. Thesis, University of Groningen, Groningen.

Wilson, R.T., Wilson, M.P. and Durkin, J.W. (1986). Breeding biology of the barn owl *Tyto alba* in central Mali. *Ibis*, **128**, 81–90.

Wilson, R.T., Wilson, M.P. and Durkin, J.W. (1987). Growth of nestling barn owls *Tyto alba* in central Mali. *Ibis*, **129**, 305–318.

Yalden, D.W., Morris, P.A. and Harper, J. (1973). Studies on the comparative ecology of some French small mammals. *Mammalia*, **37**, 257–276.

Ziesemer, F. (1980). Siedlungsdichte und Bruterfolg von Schleiereule (*Tyto alba*) in einer Probeflache vor und nach dem Anbringen von Nisthifen. *Vogelwelt*, **101**, 61–66.

Index